CLIMATE CHANGE EDUCATION

COLUMBIA UNIVERSITY EARTH INSTITUTE
SUSTAINABILITY PRIMERS

COLUMBIA UNIVERSITY EARTH INSTITUTE SUSTAINABILITY PRIMERS

The Earth Institute (EI) at Columbia University is dedicated to innovative research and education to support the emerging field of sustainability. The Columbia University Earth Institute Sustainability Primers series, published in collaboration with Columbia University Press, offers short, solutions-oriented texts for teachers and professionals that open up enlightened conversations and inform important policy debates about how to use natural science, social science, resource management, and economics to solve some of our planet's most pressing concerns, from climate change to food security. The EI Primers are brief and provocative, intended to inform and inspire a new, more sustainable generation.

Urban Climate Law: An Earth Institute Sustainability Primer, Michael Burger and Amy E. Turner

Climate Change Adaptation: An Earth Institute Sustainability Primer, Lisa Dale

Managing Environmental Conflict: An Earth Institute Sustainability Primer, Joshua D. Fisher

Sustainable Food Production: An Earth Institute Sustainability Primer, Shahid Naeem, Suzanne Lipton, and Tiff van Huysen

Climate Change Science: A Primer for Sustainable Development, John C. Mutter

Renewable Energy: A Primer for the Twenty-First Century, Bruce Usher

CLIMATE CHANGE EDUCATION

AN EARTH INSTITUTE SUSTAINABILITY PRIMER

CASSIE XU AND
RADHIKA IYENGAR

Columbia University Press *New York*

Columbia University Press
Publishers Since 1893
New York Chichester, West Sussex
cup.columbia.edu

Cataloging-in-Publication Data is available from
the Library of Congress.

ISBN 9780231202428 (hardback)
ISBN 9780231202435 (trade paperback)
ISBN 9780231554558 (ebook)

LCCN 2023012031

Printed in the United States of America

Cover design: Julia Kushnirsky
Cover image: Shutterstock

This book is dedicated to everyone who is taking action through climate change education.

CONTENTS

IV THE FUTURE OF CLIMATE EDUCATION

FOREWORD

ALEX HALLIDAY

Nelson Mandela once stated, "Education is the most powerful weapon which you can use to change the world." With people already witnessing the devastating effects of climate change, which are predicted to get far worse, we really do need to change the world—and we need climate education in order to do it. Climate scientists, known for their caution, are more often talking about a looming climate catastrophe that will be all encompassing in terms of its impacts. We also face the prospect of irreparable loss of biodiversity and the consequences of destroying the biosphere, both of which play critical roles in maintaining a healthy planet. Despite this, many parts of society seem slow to respond. Writers and filmmakers increasingly characterize humanity as sleepwalking toward disaster, as though we do not want to take seriously the reality of climate change and what it will mean for us.

Some of this reflects the deliberate spreading of misinformation and doubt about climate change.[1] It may also be because we are not yet where we need to be in terms of our understanding of the complexities of the climate system. As the climate changes, so do astonishing, unpredictable phenomena. Climate catastrophe may be on the horizon, but it is still hard to be precise about how (and how quickly) climate change will play out in specific regions, let alone in which parts of cities. This makes it difficult to convey a sense of urgency and how bad climate change disasters will be for any individual, as well as those whose well-being they carry responsibility for.

What we do know with certainty, however, is that climate change will be really, really bad for many parts of the world, and that everyone will be impacted by the knock-on effects of food insecurity, migration, and societal collapse. Providing a solid education in climate and its impacts is crucial for breaking through the lack of clarity and achieving more deliberate, focused action to decarbonize and steer society toward a more sustainable and resilient future.

The newly established Climate School is designed to harness a wide body of Columbia University expertise to deliver climate education not only to students but also to members of society more broadly. Many accept that climate change is real; they just do not know what to do about it. Climate solutions are not just about the Earth system; they also require knowledge of the social sciences, technology, health, psychology, law, architecture, ethics—to name just a few subjects. Therefore, universities are ideal for to tackle this challenge. They cover many subjects needed for delivering climate action and training well-informed leaders. They are trusted as reliable sources of expertise that are relatively unbiased by political or financial incentives.

Universities and others involved in education should not think about this as an opportunity to build an important field and establish influence. There is a moral responsibility for us to research, educate, and otherwise inform the world about climate change and sustainability more broadly. Working with society up close—whether schoolchildren, business executives, community workers, government officials, teachers, investors, nurses, or engineers—also affords academics the chance to learn.

Those most at risk, those dealing with climate change, or even just those facing questions and political challenges about it very often have a lot to teach academics. They already are confronting the issues head on and figuring out what to do. Therefore, climate education needs to be a two-way exchange. It provides a means for academics and educators to likewise improve and adapt. The future must be characterized by synergistic partnerships.

This invaluable book, written by the Climate School's foremost leaders in K–12 climate education, provides an overall framework and essential strategies going forward. It includes a review of existing resources and approaches to integrate climate into education, both formally and otherwise. It also helps us to view climate justice through the lens of education. It empowers educators to take on critical questions in classrooms and considers climate education across a wide range of settings.

Future societies need well-educated climate leaders and teachers, as well as climate-literate citizens. There is a rapid growth in jobs that require training and skills in climate and the environment. The world needs these individuals and needs them soon. This book provides a great manual for us to expand our efforts in climate education for the benefit of many people today and many more people represented by future generations, who we hope can grow up in a safer world despite an ever angrier, more dangerous climate.

ACKNOWLEDGMENTS

The authors are very grateful for the guidance provided by reviewers, colleagues, and Columbia University Press throughout the writing process.

CLIMATE CHANGE EDUCATION

INTRODUCTION

Both authors of this book work with physical and social scientists to translate their research by helping to design and implement local and international projects and studies to deliver education about climate change to broad audiences. Both authors lend their vast and diverse experiences and professional training to bring together interdisciplinary perspectives on the topic. This primer is a unique blend of efforts by the Columbia Climate School (which encompasses the Earth Institute) toward educating people about the climate and a broader overview of the field to present teaching and learning opportunities and pathways for the future.

It is important to understand the terms we have used interchangeably throughout the book, including "education for sustainable development" (ESD) and "climate change education." According to the United Nations Education, Scientific, and Cultural Organization (UNESCO), "Education for Sustainable Development (ESD) empowers learners with knowledge, skills, values and attitudes to take informed decisions and make responsible actions for environmental integrity, economic viability and a just society."[1] The international community recognizes the importance of education and training in addressing climate change. The UN Framework Convention on Climate Change, the Paris Agreement, and the associated Action for Climate Empowerment (ACE) agenda fall under the ESD umbrella. ESD positions education about climate change as a central and visible part of education. Climate action is a key priority for UNESCO's "ESD for 2030" agenda. For the purposes of this book, we discuss climate change education/education for sustainable development as the driver of climate action.

The primer provides an overview of practices that can aid in teaching and learning about earth's climate, why it has been a challenge thus far, and how we as practitioners can facilitate the important work that needs to be done. Chapter 1 discusses a systems thinking approach as the framework for addressing real-life climate-related challenges in education. The framework sets the stage for how climate education can be operationalized in various settings.

Chapter 2, on educational outcomes, provides different frameworks, including that of UNESCO and national country curricula. The chapter looks at indicators and benchmarks of learning outcomes and discusses examples from various countries. The discussions illustrate the need to not only "teach to the test" and separate physical and social sciences but to create transdisciplinary curricula that include important topics such as environmental justice. Chapter 3, on strategies on instructional design, reviews educator pedagogies and the ways climate change education can be integrated into different learning environments. This chapter gives examples of pedagogies and teaching strategies that could be adopted—a list that in no way is comprehensive.

We deliberately use "education" in the broadest sense. We believe education shouldn't be limited to classrooms and formal settings; rather, the spillover into communities and families is as significant as other forms of education. Therefore, we cover formal education in chapter 4, informal or community-based education in chapter 5, and nonformal educational settings in chapter 6. In these settings, we provide examples of tools, case studies, and practices that may be helpful for the cross-pollination of ideas.

Our review of educational case studies is by no means exhaustive but is an attempt to show the length and breadth of the ideation and implementation of climate change education in various settings. We have selected case studies that show the extensive work of researchers at the Earth Institute, the Columbia Climate School, and beyond to illustrate the broadest possible definition of ESD. We have also listed curricular resources based on our review of those that are currently available. Our motivation behind providing these case studies is to delve into the contextualization of education that must be considered before adopting any educational resources. We have highlighted the importance of communities in the various forms of education that a life-long learner is exposed to. The informal, nonformal, and formal education sections should not be seen as mutually exclusive but, rather, as complementary learning and teaching environments that learners experience at different stages in their educational pathways.

It is well known that broader hegemonic structures like conservative political systems, racial hierarchies, socioeconomic differences, and power dynamics hinder the progress of climate change education in many settings. We hope readers will view this book as an opportunity to reflect on structural inequalities that propel climate change affecting specific subsections of society in the most inequitable ways. Acknowledging social and climate injustice and reflecting on the thinking and practices in education are vital in making society more equitable and justice oriented. Therefore, we have included chapter 7, which provides a space for us to reflect on the more significant systemic issues that need to be addressed through education for a more just and equitable future. Justice and sustainability form core goals of the book, and we have attempted to use climate change education and ESD as instruments to reflect on these goals throughout the narrative.

Chapter 8 showcases the Columbia Climate School and the Earth Institute's steps to make learners future-ready through educational programs (e.g., degree efforts, professional and precollege learning, and public outreach) that look to prepare change agents to build a better and more equitable society.

This book provides readers new to ESD and climate change education with an overview of the field's current status. Researchers will be able to get a sense of how they can play a role in linking their science to education and filling the gaps that still exist, particularly in environmental justice and the piecemeal integration of climate change into classrooms. Educators could review the pedagogies and resources that may provide helpful guidance and insights into how others are working in different learning environments. Higher education institutions could look at some steps that we have taken toward formal and nonformal educational programming support. As both authors have professional experience working within Columbia University, many of our examples originate from what we have learned there. By no means do we infer that there are no better examples in the outside world; our familiarity with work done at the Earth Institute, and now the Columbia Climate School, gives us a better foundation for explaining these steps.

For young people reading the book, we remain hopeful that you will lead us to a sustainable future as agents of change backed by a systems mindset and interdisciplinary skills. We must also not forget the many community members who lead various nonprofits, religious associations, school and town green teams. This book is also for all those who have the potential to use their practices to connect to our planet. Examples of empathy toward the planet and its people will help everyone make the social-emotional connection. We have been inspired by a vibrant community whose knowledge needs to be integrated into every sphere of education and want to ensure that there is long-lasting impact on future generations.

The book brings together the authors' personal and professional experiences to make the connection to climate education and its importance. The examples and case studies reflect our opinion about the path and ambition that climate change education should take. We hope the book is helpful for readers who are very aware of the field and are seeking more insights on updated examples and case studies that they might want to try out. Some chapters also reflect our frustrations as we write about the challenges and the negligence associated with them. We hope this book will highlight the neglected areas of climate change education and urgent need to act at both individual and organizational levels.

I

WHY CLIMATE EDUCATION NEEDS SYSTEMS THINKING

1

DEFINING SYSTEMS THINKING AND CLIMATE CHANGE

We have heard the warnings. Global emissions of greenhouse gases are rising, ice sheets are melting, and our very existence on this planet is under threat. There is growing consensus around the threat of climate change. However, climate change is not merely a single-pronged problem of emissions. In reality, greenhouse gases are just one piece of a dynamic climate system that we have to consider.

The present state of our changing climate is complex. There is no silver bullet that can reverse our past actions and decisions. Addressing these complexities will require widespread change at multiple levels of society, and we cannot simply rely on siloed approaches.

"Climate change," defined as a change in regional and global climate patterns, is happening in the context of an interconnected world. Because of climate change, we need to work collectively toward achieving "sustainability," a long-term goal that will allow our planet to remain habitable for future generations. To achieve this state of sustainability, we must follow a process of "sustainable development," which is defined as the many processes and pathways that can be taken, which include sustainable agriculture and forestry, sustainable production and consumption, good government, research and technology transfer, and education and training.[1]

Another helpful way to think about the interconnections among these three concepts is that climate change may be seen as a constraint on development, but sustainable development is a key to building capacities for mitigation and adaptation, which will ultimately result in a more sustainable world. Therefore, strategies for dealing with climate change and sustainable development will be inherently complex, requiring synergy and systems thinking.

The present state of our climate did not result from a single set of factors; it was shaped by history, culture, policy, science, and economics, which have all subsequently influenced one another. We cannot change this complexity, nor should we want to. Rather than try to change or fight this complexity, we must work with it. This is what is at the heart of systems thinking.

Systems thinking is the ability to view phenomena as interconnected and dynamic; it involves understanding that the natural, social, and economic worlds are interrelated and constantly changing and that people, including oneself, are part of this dynamic system.[2] Systems thinking is widely recognized as vital to learning and understanding climate change and our Earth as a system, and we believe it should extend beyond just the physical sciences. Systems thinking is particularly useful for facilitating the complex process of sustainable development due to climate change. We need current and future populations to be equipped to conceptualize dynamic and complex problems and be able to work through transformational change and innovations to address emerging threats. In other words, if we thought of and approached climate change and sustainable development processes through systems thinking, we would not consider things in isolation. Instead, we could expand our awareness to see and understand the relationships between parts and wholes.

SYSTEMS THINKING AND THE KAB FRAMEWORK

Systems thinking has gained a lot of attention in educational settings in recent years. Therefore, it is essential to understand why these particular frameworks are helpful in facilitating development and change when it comes to preparing learners to deal with twenty-first-century climate challenges. However, using a systems thinking approach to develop and adopt a comprehensive plan for climate change education in different learning environments still faces an uphill battle.

In the United States, climate change as a science and a process has become a highly politicized issue. There are those who outright deny that it is happening. Then there are those who acquiesce to the data but question the role of human beings in the process—because, after all, our climate has been changing naturally throughout our history and will continue to do so. Finally, there are those who do believe that climate change is happening, but this knowledge has not translated into any behavioral change.

There are a lot of driving forces behind why climate change is still such a contested topic, particularly when it comes to whether we should be teaching it. Up to now, many attempts to educate people about climate change have relied on scare tactics that generate fear rather than emphasize solutions-oriented thinking. That approach can backfire and inhibit learners. They know it is happening, they have the knowledge, and yet their attitudes and behaviors have not shifted. Here is one potential perspective on why that may be happening.

The knowledge, attitudes, and behavior (KAB) approach is a helpful framework for understanding why simply knowing about climate change is just not enough. KAB has been widely used in public health and medicine, as well as to evaluate organizational performance. However, there are important linkages and lessons we can borrow from this approach to understand the importance of interactions and components of how we learn and make decisions based on our acquired knowledge.

KAB posits that learning must go beyond the narrow-mindedness of knowledge-based development because knowledge is only a small portion of what we want future generations to learn. The notion that learning is more than just knowledge is not new.

As early as 1956, Bloom began developing a taxonomy of instructional objectives in three domains: cognitive, affective, and psychomotor.[3] Since then, research has confirmed the importance of these constructs as outcomes of learning, and strong ties have been found between the cognitive and affective attributes of the learner and their impact on the acquisition and comprehension of information.[4] Icek Ajzen and Martin Fishbein[5] further reported that while it is not the sole indicator, attitude is a factor in determining behavior, and Min-Sun Kim and John Hunter[6] added that the higher the attitudinal relevance, the stronger the relation between attitude and behavior.

Since those studies, educational researchers have gone on to adapt Bloom's taxonomy into a construct for teaching, learning, and assessment that looks at not only knowledge but also attitudes and behavioral changes.[7] Other researchers have since proposed an assessment approach for learning that measures knowledge gains, the heightening of learner attitudes, and the impact of knowledge and attitude on behavioral change.[8] To further understand the KAB approach, let us break down each of its components.

1. Knowledge—Within a domain, knowledge embodies all information that a person possesses or accrues related to a particular

field of study.[9] We can acquire knowledge in a variety of ways through problem-solving environments.[10]

2. Attitude—Traditionally, attitude has been broken down into two big ideas: behavioral and cognitive. Allport[11] defines attitude in a behavioral sense, as a mental state of readiness conditioned by stimuli directing an individual's response to all objects related to it. In contrast, Thurstone[12] assumes that an attitude is the effect for or against a psychological object rather than a behavioral object, as others suggested.
3. Behavior—Behavior is the way in which a person, organism, or group responds to a certain set of conditions.[13]

While each of the KAB components can be difficult to measure objectively, what is more important for us to take away is how each of these components interacts (in a dynamic way) to further learning. The relationship between these three components is complex, so what an individual knows may inform their attitude about that topic but may or may not affect how they behave as a result. Alternatively, knowledge and attitude are not necessarily predictors of behavior. Therefore, the main essence of KAB is that the relationship between these three dimensions is dynamic and reciprocal; they interact in a nonlinear way and have important connections to one another.[14]

Shaping people's attitudes through social-emotional learning skills helps them form deeper connections with others and with nature. Some educational systems have begun to reorient themselves toward a broader sense of purpose that aims to go beyond developing the cognitive, academic, and technical competencies of learners and also aims to promote learner well-being by developing social and emotional competencies. For example, learning is grounded in the self-awareness of actions, self-management of thoughts, and social awareness of contextual realities; these are key areas for forming relationships and encouraging responsible decision-making.

Let's apply KAB to climate change education and understanding. KAB helps us understand why gaining knowledge about climate change has not necessarily led to equal amounts of action and behavioral change. In other words, just because you have knowledge about something (e.g., why reducing single-use plastics is good) does not automatically lead you to take an action that is useful (e.g., continuing to choose not to recycle).

Another potential way to think about this disconnect between knowledge and action through the KAB perspective is that people are aware that climate change is happening (the knowledge), but it has not yet affected them personally (the attitude nor the behavior). For example, a parent may think about climate change (which may seem like something that was happening very far away and at a global scale), but it does not play a role in their daily activities. So it's quite possible for someone to be familiar with climate change, but the impacts on a daily basis may not be as clear, so that knowledge and awareness of climate change do not result in shifts in attitude or behavior.

Often, it is difficult to convey and connect a complex topic like climate change to what is happening in our daily lives. Someone may know that average global temperatures are rising each year but not realize that the reason the price of their groceries keeps going up is that farmers are dealing with more drought and uncertainties during their growing seasons.

Someone may also know that there is a tendency for their neighborhood to flood during heavy downpours but not realize that with rising global temperatures comes more rain and that the warming of oceans results in more hurricanes. That is because higher ocean levels and air temperatures increase the possibility for evaporation and cloud formation (and at higher temperatures, the air holds more moisture, which leads to an increase in precipitation intensity).

KAB can help us illustrate that what we know, how we think, and what we choose to do interacts in dynamic ways and is not always logical. This insight is useful because it can help us explain why the idea of climate change may be gaining traction, but we are still at odds and debating whether we should be teaching climate change content in different learning environments.

MOVING BEYOND KAB

Rather than continuing to push specifically for teaching future generations only about climate change and the scientific knowledge behind it, we believe efforts should focus on implementing integrated and holistic education around earth systems science through systems thinking. Specifically, how do we bring the spheres of the Earth (i.e., ocean, atmosphere, biosphere, hydrosphere, and geosphere) into a framework that helps learners understand our environment and how all these pieces are interconnected. And ultimately, how the myriad feedbacks and balances within this framework are critical for understanding and do something about our changing climate.

We cannot use a hardwired approach to climate change education. We cannot simply assume that just because a learner has gained content knowledge about how their climate is changing, they will then continue in a linear way to change their attitudes and behaviors.

Learning is inherently a nonlinear process, so expecting students to progress through a hierarchical pyramid can be misleading. Knowledge absolutely matters, and there are learning situations in which foundational knowledge is probably the most important skill. However, with a complex topic like climate change, which is really a systems problem, we have to push for other ways of learning and teaching beyond conveying knowledge.

Because climate change itself is not a linear process, we should not be learning and teaching it in a linear way. Learning about climate change may start with knowledge, but it should not stop there. We need to be providing learners of all ages with opportunities to create and analyze based on knowledge they have gained. While they're doing so, they will also continue building knowledge and understanding. Through a process of using their minds and hands actively to create and analyze their understanding, individually and collectively, learning becomes dynamic, synergistic, and interrelated. In other words, it becomes a systems process in which all the pieces are deeply connected to one another, and it is never static.

To move away from siloed teaching of climate change, we should emphasize learning about this complex challenge through systems thinking. We need to change how we think of climate science education and make it more project-based and more connected to the lived realities of students.

Teaching and learning through a systems thinking lens is truly interdisciplinary, and it is so much more than just science education. Because systems are intricately complex, there are many ways to study, learn, and teach them.

A fundamental understanding of the systems that make up our planet is composed of important components, for example:

1. Natural processes that influence our climate systems at different scales;
2. anthropogenic processes (e.g., the role of human beings);
3. the multifaceted impacts that changes in climate have on other systems (e.g., physical, economic, social, geographical);

4. how we now have evidence and data across all systems to demonstrate climate change is under way;
5. the range of effects, risks, and uncertainties brought on by climate change and strategies of adaptation, mitigation, and resilience;
6. social justice and uneven impacts of climate change on communities; and
7. community engagement, environmental stewardship, and climate action.

These interrelated and connected ideas, or systems, should make up the foundation for climate education efforts. Implementation and delivery of ideas may differ, but these important components serve as a helpful jumping-off point for future education and training programs that aim to increase climate literacy for future learners. The next chapter discusses outcomes, indicators, and skills that we expect from students when a systems thinking lens is applied.

Efforts to bring climate science and climate change into school curricula have been controversial at best. In 2008, following a major effort by numerous groups of scientists and other stakeholders to educate the public about climate change, 72 percent of Americans believed that climate change was happening.[15] By 2010, however, this figure had dropped to 50 percent, and the percentage of those who actively did not believe in climate change had doubled, from 17 to 36 percent.[16]

Systems thinking can be an important stepping stone that may lead general populations to move toward greater understanding and consideration of our planet as a system. Through the introduction of systems thinking in our education and training efforts, we may be able to facilitate a more ecological worldview that transcends political ideologies and party affiliations. Strengthening our systems thinking in all learning environments could serve as a useful strategy to help move away from the debate about whether climate change and global warming are real and toward attitudes and mindsets that place value on our natural world and why and how it needs to be preserved and protected.

This wide-reaching perspective of systems thinking should form the foundation of educational efforts to promote both an understanding of climate change and valuable attitudes and behaviors. To address complex environmental challenges, we need learners to be able to conceptualize dynamic, complex, and interconnected systems, because only then can we bring about system-level changes that will tackle climate change.

In particular, a shift toward systems thinking in our education system means a shift away from the scare tactics approach to teaching future generations about climate change. We are not saying that every school should have a stand-alone climate change class that all students are required to take. Instead, we want to see the interconnected processes that make up earth systems woven into all subject areas.

A helpful historical reference here is the Bretherton diagram (see figure 1.1). This first appeared in the Bretherton report, *Earth System Science: A Closer View*, published by NASA in 1988. It outlined two critical principles: (1) that all earth processes and phenomena can be understood as dynamic systems that transform or transport matter and energy in accordance with the laws of chemistry and physics and (2) that all these earth systems are interconnected, so that no system can be understood in isolation from any other.[17]

The report's findings are a strong reminder about why it is impossible to consider climate science in isolation and expect people to understand it, embrace it, and change their attitudes and behaviors toward it. It makes sense only as part of an integrated, holistic education in systems science.[18]

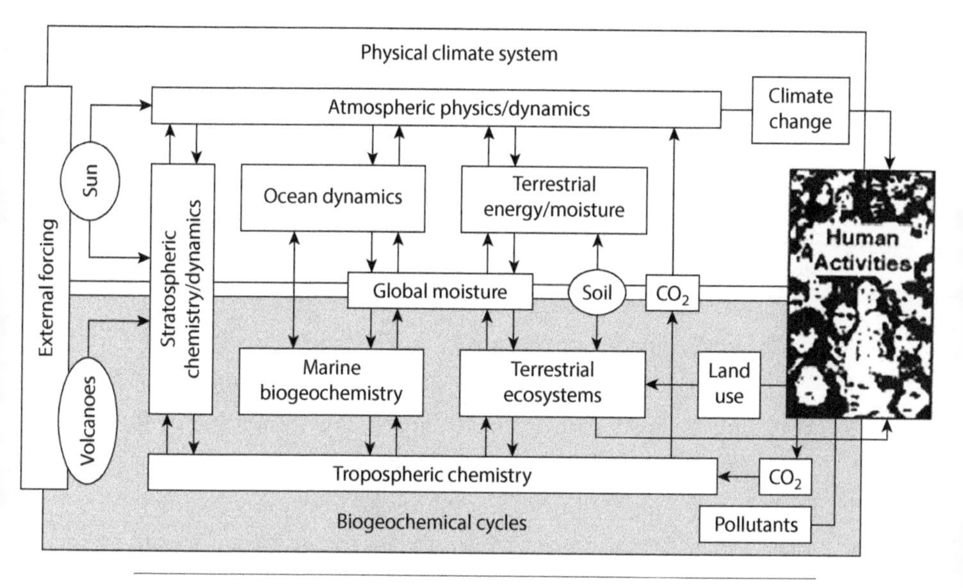

FIGURE 1.1 Bretherton diagram

Source: Earth System Science-Overview, NASA. 1988, 19.

If we are to prepare future generations for jobs that do not yet exist and for risks and uncertainties we cannot accurately predict, we need to educate them using systems science so they can understand what is going on and be able to connect climate content to larger systems at play.

Climate change as a stand-alone subject remains controversial. We have seen fierce debates about why schools should or should not include climate science as part of their standards and curricula. Adding climate change education on top of everything else that is already in the curriculum is likely to result in some teachers doing a great job and others either not believing in it and therefore not doing it at all.[19] Basic understanding of climate systems should be integrated into subjects such math, science, social science, and English either through activities or project-based learning tools. In this way, climate change education would not be treated as an add-on but included as an integral part of teaching and learning cross-cutting subjects.

We need to fundamentally change the approach toward teaching climate change. That is the only way we can prepare future learners to gain the capacity to make evidence-based decisions based on a fundamental understanding of the processes of earth systems, how those systems are impacted, and the relationships between those systems locally and globally.

A dynamic earth systems education grounded in systems thinking should not replace other educational frameworks and standards. It should complement existing structures, be integrative, and emphasize both physical and social systems. Systems thinking is also not a topic or class that students can opt in to study. Instead, it should be treated as a holistic learning approach in which learners can acquire skills such as evaluating evidence, projecting risks and uncertainties, weighing options based on the information in front of them, and using their critical thinking and application skills to make well-reasoned decisions.

To facilitate a move toward systems thinking in our pedagogy, we almost need to work backwards. First, let's think about the bigger picture of what the challenges young people are going to face when they graduate and then map out priorities and educational activities that will prepare them to address those challenges. If future employers want employees to be comfortable with being uncomfortable, have critical thinking skills, and possess the ability to envision and analyze what is not there, then we need to make sure that what students are currently learning goes beyond yes or no answers. We need to emphasize that learning and teaching have multiple answers and to ensure that learning is connected to real-world problems.

EXAMPLES OF SYSTEMS THINKING AND CLIMATE EDUCATION

Examples exist of how this integration of systems thinking into climate change is already being done, particularly through education for sustainable development (ESD) and the Next Generation Science Standards (NGSS). ESD encompasses a multifaceted and systems thinking approach globally to teaching and learning about our world that is crucial for preparing future generations to tackle complex global challenges such as climate change. Traditionally, education in this realm has focused on environmental education, which has often been limited to teaching about the environment predominantly through knowledge acquisition. However, a new approach is needed for a new and more complex set of problems, specifically for climate change.

ESD, a framework put forward by the United Nations, is an example of a more active approach to learning about all issues that encourages critical thinking, social critique, and analyses of multiple contexts and systems.[20] It also involves discussion, analysis, and application of multiple values. ESD pedagogies often draw on both the physical and social sciences to stimulate creativity and interdisciplinary thinking. It also emphasizes the role that individuals have in creating change that leads us all toward a more sustainable future.

The ESD literature relies on the whole-school approach to ensure that ESD is taken up on a schoolwide level with all the relevant policies rather than being limited to curricula taught in classrooms.[21] It also involves teacher professional development, routines and structures, the role of the principal, student leadership, campus sustainability, and other ideas. The whole-school approach has been extended to a whole-institution approach, which involves school governance, facilities and operations, community partnerships, and teaching and learning as the four pillars, along with their associated policies and activities.[22]

The K-12 Next Generation Science Standards (NGSS) in the United States is another example of systems thinking within educational pedagogy. NGSS emphasizes crosscutting concepts as one of the core dimensions of scientific learning. It represents a vision of science and engineering learning designed to bring these subjects alive for all learners by emphasizing the joy and satisfaction that comes from pursuing compelling questions through a process of discovery. The research on learning science and engineering that informed NGSS emphasizes that science and engineering involve both knowing and doing; that developing rich, conceptual understanding is more productive for future learning than simply memorizing discrete facts; and that those learning experiences should be designed with coherent progressions over multiple years in mind.[23]

Both ESD and NGSS are helpful frameworks that can guide our understanding of how we move from siloed learning and teaching toward systems learning and teaching in different educational environments. The topic of climate change and how we go about teaching and learning of this subject alongside existing curricula is a relatively new concept, and how it has been adopted has varied greatly in many countries. If we are to have any hope of achieving sustainable development and ensuring that future generations can enjoy the same habitable planet that we currently do, we need to shift away from single-discipline ways of operating toward multidisciplinary approaches that allow for the recognized complexity of and uncertainty within systems.[24]

In the next chapter, we will look at the skills and outcomes that we aim to achieve for learners through a holistic approach to earth systems education and thinking.

II

CLIMATE CHANGE EDUCATION AND FUTURE WORKFORCES

2

SYSTEMS THINKING SKILLS AND OUTCOMES

In this chapter, we take a closer look at outcomes in a learning environment where an integrated systems thinking approach to climate change education is used. Questions that are important to consider in this context include these: What should students be learning at each grade level, and what might they be expected to know (skills) for their future transdisciplinary jobs? What competencies (knowledge and behaviors) are developed as a result of of integrating systems thinking into the curriculum, and how will we measure their success? The chapter also uses literature-driven frameworks to define the skills and attitudes that are aligned toward achieving climate literacy.

Competencies are combinations of attitudes, skills, and knowledge that become the guiding principles of teaching and learning for school, work, and life. Key competencies help establish the big areas that need to be covered, forming the basis of what content helps achieve success in those areas. Key competencies are more likely to be achieved if those in teaching roles see them as part of the curriculum, which becomes the foundation for the learning opportunities that teachers should provide.[1] How cognitive competency interacts with behavioral competency is the key to whether or not knowledge gets translated into action. For example, if the knowledge component is climate literacy, how it interacts with skills such as critical thinking and attitudes such as empathy and proactiveness will help make environmental action possible in the neighborhood.

Who is a climate-literate person? According to Global Change Research Program, a climate-literate person has the following abilities:

- understands the essential principles of Earth's climate system,
- knows how to assess scientifically credible information about climate change, and
- communicates about climate and climate change in a meaningful way and makes informed and responsible decisions about actions that may affect the environment.[2]

Christina Kwauk and Rebecca Winthrop define climate literacy as being able to "map and monitor local environmental challenges, analyze local practices, policies, and laws that perpetuate or enable these challenges, and design and implement or advocate for a sustainability plan that addresses the root cause(s)."[3] To achieve the transformative skills that Kwauk and Amanda Braga describe, climate change education needs to meet certain conditions.[4] First, climate crises cannot be taught without consideration of social crises, and thus justice and equity are integral concepts. Second, climate education needs to be introduced across all subjects, not exclusively as a separate subject offered as an elective. Third, climate education aims to improve knowledge on the topic so learners can act on it. Therefore, civic responsibility, advocacy, and communications are included as elements of climate education. (Each of these concepts is explained later.) Fourth, social-emotional learning (SEL) has become a critical part of learning, especially in light of the COVID-19 pandemic, and is much needed in a comprehensive climate change education curriculum.

In the international context, climate literacy varies greatly. There are often many challenges beyond curricula, including access to education and basic literacy. In many low- and middle-income countries, children cannot read at grade level.[5] Thus, adding a layer of environmental awareness seems more a dream than a realistic goal. Moreover, the learning gap has increased since COVID-19. World Bank estimates show that 53 percent of the children in low- and middle-income countries experience learning poverty; that is, they are unable to read or understand a simple textbook by age ten.[6] Therefore, the task at hand is huge: not only must millions of children attain basic competencies in reading and writing, but they also need to quickly become climate literate.

Climate change education is sometimes given lip service and offered without integrating into the curriculum and teaching the context and local environmental needs of a particular place. It's treated as a "good add-on" but not essential core material. If we are to have a chance to bring systems thinking into how we educate learners about climate change, we need to significantly rework the current practice both from a top-down (from state or national frameworks and policies) and a bottom-up approach (classroom-level change).

We should be asking two important questions as we think about how we might approach this change. What should a (for example) grade two student need to know about systems through the various subjects offered in schools? How do we define what a climate-literate person is for each grade? These questions do not necessarily have a simple answer. Grade-wise competencies are not established globally, and very rightly so. Each country has its own national curricular policies that translate to the standards it adopts, which are then adapted or adopted by each subnational entity or state. These standards are then aligned to the textbooks and content that the schools teach.

In some countries, like India, sustainability is taught as a separate subject under environmental sciences for primary grades (one through five). In upper grades, it is included in social sciences and sciences. In the United States, every state has its own mandates on climate change education. In the United States, in New Jersey, for example, sustainability is not included as a separate subject; the state government has mandated that climate education and sustainability need to be integrated into all subjects starting from kindergarten. Environmental science is also offered as an elective in New Jersey high schools.

Internationally, to provide guidance and push for the integration of systems thinking into learning environments, useful frameworks are suggested by the Organization for Economic Cooperation and Development (OECD), UNESCO, the Brookings Institution, and others to advise countries on the competencies that students should master. Some of these frameworks are outlined in this chapter. These frameworks assist in providing global consensus on school-age students' competencies and learning outcomes. They help answer the question, irrespective of the content, "What are learners expected to acquire in terms of knowledge and behaviors in general?"

Educators can use these frameworks in several ways. First, we can review their objectives and gain insights into how competency mapping is done. As the world around us evolves, new skills are needed to cope with uncertainty. COVID-19 taught us the importance of SEL skills. Therefore, looking at competency mapping and making connections between SEL and climate literacy may help students navigate those uncertainties. Second, the frameworks can provide a macro overview of what students can master. We can also use the frameworks to better understand how educator professional development workshops can become strategic places to reflect on the larger goals defined in the frameworks and reconnect to bigger-picture goals for classrooms, schools, and/or districts. Reflecting on such frameworks collaboratively with educator colleagues can enable them to find different pathways to meet the defined educational goals. Third, frameworks could provide sample lesson ideas and plans. As discussed earlier, systems thinking requires a transdisciplinary approach to teaching. For example, a lesson on English vocabulary about energy that introduces the concept of renewable energy sources can help students understand energy policies across states, connect to sustainable development goals (SDGs, discussed later), and encourage them to create a science model on solar energy. To integrate these various subjects and ideas and successfully deliver a lesson requires a fundamental understanding of a transdisciplinary framework. Fourth, the connections between frameworks and country/subnational curricular standards may lead to useful discussions that result in development of new lessons and serve as a strong complement to national or subnational curricular standards.

Let us now look at an example of a theoretical framework to understand the various dimensions of ESD (figure 2.1). This framework aids in parsing out the components of systems thinking approaches and links to the global sustainable development goals. Such a theoretical framework is beneficial in combining the various aspects of sustainability as suggested by the United Nations. It keeps sustainability and justice at the center, aligned to ESD and global citizenship education. The remaining thematic areas are as suggested in the United Nations SDG 4.7. The entry points to these types of education are through schools, families, and communities. The framework is

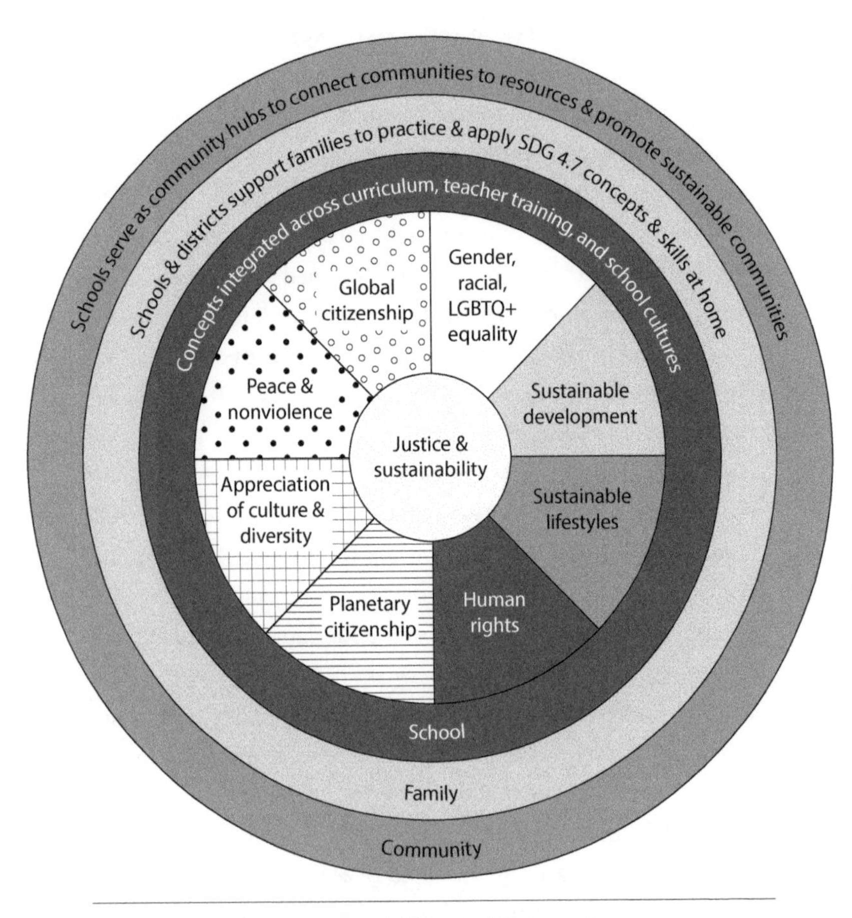

FIGURE 2.1 Theoretical framework

Source: Mission 4.7, "Planning and Policy for Transformative Education," https://sites.google.com/view/mission47/home/policymakers-guide; created by Radhika Iyengar and Tara Stafford Ocansey.

one way of approaching systems thinking and understanding the various components (human rights, planetary citizenship, peace education, and others) that educators could address through their lesson plans and teaching. The components could tie to the knowledge, skills, and attitudes required to promote climate literacy and action.

It is important to note that all the components are interconnected and overlap to reinforce the systems thinking approach. Educators could use this framework to incorporate these components into their lesson plans and make them interdisciplinary. It is also a good reminder that contextual relevance is important. Connecting to the learner's immediate environment through schools, families, and communities will help make practical connections with the learning content.

Two elements are core to this theoretical framework: justice and sustainability. In the next two sections, we discuss these concepts and their core competencies.

JUSTICE AND ITS CORE COMPETENCIES

There is a need to recognize that intersectionality exists in teaching competencies. Justice is a multidimensional topic, which makes it harder to align it with competencies, and thus often it is underrepresented in learning environments and curricula. For example, a key lesson from the COVID-19 pandemic is the intersectionality of climate, social (race), health, and economic factors that have affected everyone. Therefore, these linkages must be recognized and brought into education as related issues across all subject areas. Kwauk and Casey also raise an important topic: that gender-blind focus on competencies is seriously flawed.[7] It also undermines a lot of existing literature on ecofeminism, which has created considerable activism in communities worldwide. The authors also emphasize that marginalized communities such as individuals with disabilities, indigenous groups, those who identify as LGBTQI, underrepresented minorities in science, refugees, and displaced peoples are often left out of educational content, curricula, and texts. Therefore, place-based and people-centric approaches need to be adopted to translate these skills in schools and communities.

Intersectionality aligns to global citizenship education with civic action, ethics, and justice, and care for the people and the planet ought to be addressed in curricula, which is why justice issues are key to educating about climate change. Kwauk and Olivia Casey suggest that the green new learning agenda needs to go beyond climate education and action.[8] It also needs to address intersectional topics of social justice, equity, gender, and economic and social marginalization to highlight the cross-linkages with climate justice issues and the relevant action. Needless to say, science and technology will not bring transformative change in the Anthropocene.

Learners also need to be inclusive in their thinking and practice. They must realize that humans coexist on the planet with other beings, and everyone must coexist peacefully. Respecting the rights of fellow beings as Earth's resources grow scarce has become an unfortunate reality. Restoring the balance through mutual cooperation will require everyone's concerted effort. Thus, preparing the citizens of the world to adapt and become resilient by taking care of the most marginalized should be a climate education priority.

COMPETENCIES, SKILLS, AND GOALS FOR SUSTAINABILITY

Sustainable Development Goal 4

The United Nations Sustainable Development Goals—specifically goal 4—is another helpful framework for looking at learning

TABLE 2.1 UN SUSTAINABLE DEVELOPMENT GOAL 4

Goal 4. Ensure inclusive and equitable quality education and promote lifelong learning opportunities for all.	
Target 4.7 By 2030, ensure that all learners acquire the knowledge and skills needed to promote sustainable development, including, among others, through education for sustainable development and sustainable lifestyles, human rights, gender equality, promotion of a culture of peace and nonviolence, global citizenship, and appreciation of cultural diversity and culture's contribution to sustainable development.	**Indicator 4.7.1** Extent to which (i) global citizenship education and (ii) education for sustainable development, including gender equality and human rights, are mainstreamed at all levels in (a) national education policies, (b) curricula, (c) teacher education, and (d) student assessment.

outcomes that yield a systems approach to climate and earth systems education (table 2.1). SDG 4 helps to address the question, "What are the learning goals for all school-age students?" It has also given the world an opportunity to think about basic literacy and numeracy and the possibility of defining what education goals should be for us to move toward a truly integrated approach to education for sustainable development.

The list of all SDG 4 indicators is a significant departure from the previous Millennium Development Goals,[9] in that it moves away from the compartmentalization of subjects and content areas in favor of education that allows for seamless learning across subjects through experiential and inquiry-based education. The standards outlined in SDG 4 provide a list of competencies (what students need to achieve after graduating from the grade). These competencies are then matched to units or topics the students must master. The units are aligned to students' expected cognitive skills. Lesson plans have been developed that tap into cognitive skills that aid in acquiring knowledge and affective skills such as attitudes and behaviors and can produce the desired competencies.

All the targets in SDG 4 help us understand that educational possibilities are not siloed but are deeply connected as a system. Basic numeracy and literacy are the building blocks of higher-order thinking. With SDG 4, the focus on achieving learning outcomes and expanding learning opportunities throughout one's life span became the center of education dialogues. Within the ten SDG 4 targets, we see the focus on the intersectionality of all aspects of education, which encompasses not just educational content of curricula but also teacher education, equity and justice, assessments, and much more (figure 2.2). Above all, this target emphasizes learning and teaching that will lead students to understand issues that directly relate to them, their society, and the rest of the world. It also highlights the linkages between their lived experiences and global processes.

However, implementing educational programs aligned with SDG 4 targets is easier said than done; each country and state has learning standards that are often subject- and grade-specific. There is also a significant variance between developing and developed nations. In most developing countries, the reality is that by the time children reach first grade, they are developmentally behind in every learning metric.[10] Entering grade one should be the stepping stone to success, but many schools fail to make children even grade one ready. The learning lag continues to widen further in primary education. Even among the children who attend school, 130 million of them cannot read or do basic math after four years. In assessments implemented by ASER/UWEZO in Ghana and Malawi, more than 80 percent

FIGURE 2.2 Elements of justice and sustainability

Source: Mission 4.7, "Planning and Policy for Transformative Education;" created by Radhika Iyengar and Tara Stafford Ocansey.

of students at the end of grade two could not read a single familiar word such as "the" or "cat."[11]

Despite the challenges, the targets in SDG 4 have helped to standardize learning goals, but at the same time, lesson planning and adaptation remain a very inflexible process. Many debate the learning standards themselves. However, some common strategies are generally agreed on by the learning science community (i.e., project-based and experiential learning), and those strategies are aligned with ESD and systems thinking goals.

ESD is also a helpful framework to consider in this context. It empowers learners to make informed decisions and take responsible actions for environmental integrity, economic viability, and a just society for present and future generations while respecting cultural diversity. It is about lifelong learning and is an integral part of a systems approach to climate education. UNESCO's education for sustainable development framework helps to operationalize this lifelong learning approach.

UNESCO's Learning Framework on Sustainable Development

After the SDGs were formalized in 2015, keeping SDG 4.7 as the focus, UNESCO provided a list of competencies.[12] UNESCO's learning framework, published in 2018, provides competencies that are linked to each of the seventeen SDGs. For each SDG, the competencies are divided into three categories: knowledge or cognitive, social-emotional learning, and behavioral. This framework is particularly useful for teaching about sustainability with the sustainable development goals as the entry point. Although the competencies are not linked to grade level, they can easily be matched to grade-level competencies in any country and customized for grade-level teaching. Let us dig deeper with examples and interpretation of each of the main domain areas listed in the UNESCO framework: knowledge or cognitive, social-emotional learning, and behavioral.

UNESCO's learning framework emphasizes the importance of gaining knowledge about the SDGs. It takes each goal and explores the knowledge area using science. The cognitive content area draws from all subjects and topics linked to sustainability.

A similar process is adopted by New Jersey, which has taken New Jersey Student Learning Standards for Science (NJSLS-S) and cross-linked them to climate change topics. "Standards in action" is a very apt title for recommending the actions relating to social studies that are subject to climate change. The connection to climate change is made through the pedagogical approaches of "developing the requisite skills—information gathering and analysis, inquiry and critical thinking, communication, data analysis, and the appropriate use of technology and media—at the youngest grades for the purpose of actively engaging with complex problems and learning how to take action in appropriate ways to confront persistent dilemmas and address global issues." Therefore, the state takes the existing standard for science and maps the relevant climate change content area to the relevant standard to integrate climate education into the curriculum. This content integration is essential for cutting across subject areas for a holistic treatment of any sustainability topic.

UNESCO's learning framework on sustainable development also lists social and emotional learning as an integral part of students' learning. The following section discusses SEL and its interconnections with sustainability.

Social-Emotional Learning and Skills and Greening the Future

COVID-19 has taught us that SEL needs to be addressed in all subjects and all types of learning. With SEL being the immediate need in curricula across all levels, values such as empathy toward others and the planet will help communities recover from this pandemic and avoid future events.[13] Religious leaders like the Dalai Lama and Pope Francis have emphasized empathy toward people and the environment. The importance of value-driven education has been stated before. The first word revealed to the Prophet Mohammed (peace be upon him) was *Iqra*, meaning "to read" (emphasis on seeking knowledge and learning). Pope Francis's *Laudato Si'* item 4, or "On Care for Our Common Home," urges us to empathize with our environment. This environmental encyclical is a meeting point between the environment and spirituality. He delicately uses scientific words such as "global warming" and "carbon emission" and frames them in a spiritual perspective. Pope Francis takes his inspiration from St Francis of Assisi and relates to nature as "sister earth," "brother sun," and "sister moon." He urges us to connect with different aspects of the planet to cultivate the "ecological virtues."

A broadened understanding of SEL incorporates empathy for our shared home on Earth as an extension of empathy for one other. SEL links individual and community resilience to environmental resilience, which can help raise awareness of how issues like environmental degradation and biodiversity loss pave the way for the spread of deadly pandemics like COVID-19, droughts that cause mass hunger, and other human challenges.

Pope Francis calls for a "consciousness-raising" to further prevent all the health and environmental risks caused by humankind. An approach to SEL that incorporates empathy for people and the environment will help us be mindful of our actions and look deeper within ourselves to break the "myths" of a modernity grounded in a utilitarian mindset (individualism, unlimited progress, competition, consumerism, and the unregulated market). This reflective practice will also aid in "establishing harmony within ourselves, with others, with nature and other living creatures, and with God."

Harm to our environment has taught us about empathy in real life; how can we take this lesson and integrate it into our schooling systems? Pope Francis thus explains that the real purpose of environmental education, which can incorporate SEL, is not to teach facts. He is suggesting an approach to questioning one's practices and meaning-making as a way to teach about the environment. He urges educators to encourage "ecological ethics" in developing "ecological citizenship." The pope gives examples of small but essential practices from which we could all learn through this form of education, "such as avoiding the use of plastic and paper, reducing water consumption, separating refuse, cooking only what can reasonably be consumed, showing care for other living beings, using public transport or carpooling, planting trees, turning off unnecessary lights, or any number of other practices." This could be such a profound way of "cultivating sound virtues" whereby people will be empowered to "make a selfless ecological commitment." Therefore, connecting with the environment requires connecting with oneself first.

UNESCO's ESD learning framework also uses behavior change as a critical domain area emphasizing that knowledge acquisition should lead to civic environmental action.

CHANGING BEHAVIORS THROUGH CIVIC ACTION

UNESCO's ESD learning framework helps us realize that cognitive knowledge should be translated into action so it can have an impact on society. Civic education is usually associated with a social science curriculum that is geared toward making society more socially conscious. Rebecca Winthrop calls for a focus on civic action as part of twenty-first-century school curricula.[14] The author associates civic action with a functional democracy.

For instance, New Jersey's social science standards have outlined the proximal learning outcomes as follows:

- Is civic-minded, globally aware, and socially responsible;
- Exemplifies fundamental values of democracy and human rights through active participation in local, state, national, and global communities;
- Makes informed decisions about local, state, national, and global events based on inquiry and analysis;
- Considers multiple perspectives, values diversity, and promotes cultural understanding;
- Recognizes the relationships between people, places, and resources as well as the implications of an interconnected global economy;
- Applies an understanding of critical media literacy skills when utilizing technology to learn, communicate, and collaborate with diverse people around the world; and
- Discerns fact from falsehood and critically analyzes information for validity and relevance.[15]

These outcomes are geared toward the state's larger mission: "Social studies education provides learners with the knowledge, skills, attitudes, and perspectives needed to become active, informed, and contributing members of local, state, national, and global communities." In the New Jersey case, the standards help provide the space and motivation for value and attitude formation, local action, and behavior change. The curricular standards also connect global knowledge to local action.

In conclusion, in this chapter, we began to address what students must learn through the systems thinking approach. Our aim was to provide a foundation for discussing broad goals and learning objectives by using the competencies in a variety of frameworks to drive the content taught in schools. The goal was to provide a bigger-picture thinking space to understand what students should learn to advance a more sustainable world. Next, we explore specific strategies in instructional practices to look at how we can implement goals around skills and outcomes and turn them into action.

3

STRATEGIES IN INSTRUCTIONAL DESIGN

We experience learning every single day, whether we realize it or not. While the place of learning may vary (e.g., a classroom, the workplace), content (and the ways in which it is shared) influences the ways we learn and from early on and throughout our lives.

Therefore, thoughtful curriculum and instructional design plays an important role in educational programming effort toward addressing change at scale. Curriculum and instruction are the pillars of how we educate, train, and engage learners in formal and informal learning environments. Educators play a critical role in ensuring that thoughtful curriculum can be delivered in a meaningful and sensible way so learners grasp content and gain skills. Next, we will explore how a thoughtful curriculum, coupled with prepared educators, can ensure that learners walk away from different learning environments with important outcomes and skills.

CURRICULUM AND INSTRUCTION

We believe that curriculum and instruction around the concept of systems thinking need to be part of education and training efforts moving forward. Over time, educational experts have carefully studied criteria that lead to effective curriculum design. These factors are important to consider when we think about how we may update or introduce climate change content through the lens of systems thinking into different learning environments.

In the previous chapter, we outlined the importance of moving toward curriculum design that encompasses a systems approach to learning about climate change—in other words, that systems-oriented education is the outcome and that learners end up seeing the big picture. It is a holistic plan and a macro-level view (i.e., the "what" we need to learn) of how we might piece together strategies, materials, and experiences to meet that end goal.

In this chapter, we will focus specifically on a few instructional design components that we believe should be in place to meet the curriculum goals of climate change education. Similarly important, we see instructional design as the micro-level processes, tasks, and materials that are needed to meet curriculum design goals. In other words, it is the "how" part of teaching what we want students to learn.

Before we dive into the instructional design components, it is important to highlight the skills and outcomes that are important to consider when designing curricular materials and instructional content. Working backward and thinking about what we want learners to walk away with is a critical first step. Next, we look at work that has been done in the area of "green" skills and what that means for a sustainable future.

THE GLOBAL NEED: GREENING THE TWENTY-FIRST-CENTURY SKILLS FOR A SUSTAINABLE FUTURE

The Intergovernmental Panel on Climate Change released an alarming report that discusses the disastrous impact of global warming of 1.5° C above pre-industrial levels.[1] Without increased and urgent mitigation to sharply decrease the greenhouse gas emissions, irreversible loss and more climate crises are inevitable. Education's untapped potential needs to be put to work here. To have pro-climate changes at the personal, political, and practical levels, green skills for that transformation must be developed. This development will ensure that beliefs, values, worldviews, and paradigms interact with systems and structures to influence behaviors and technical responses to climate change. Kwauk and Casey[2] categorize these green skills into three "buckets" (figure 3.1). First, skills for green jobs are those required for green jobs of the future. These jobs are in sectors that produce low levels of carbon emissions and promote a green economy. They are, as the authors call them, "instrumental skills" that are job focused. These jobs are essential for the functioning of a green economy.[3] Second are green life skills, which cross-cut SEL, cognitive skills and are adaptive and transformative. In the third bucket are green skills for transformation, which have a postmodern perspective with more of a rights and justice oriented lens. These skills have transformative capabilities.

Note that this list is not exclusive; rather, the idea here is to present some skills that hint at the categories of skills that fall into each bucket. They should be considered essential input that will be matched by knowledge, attitudes, and behavior to lead to the desired competencies.[4]

Green skills require more interdisciplinary and collaborative crisis management education than is currently taught. Skills that create a regenerative economy will have to combine earth sciences and social sciences with technological advancements. Ultimately, systems literacy is going to be key. Our changing climate is the greatest environmental threat of the century. Regardless of

Skills for Green Jobs

Skills aimed at fulfilling the requirements of green jobs and supporting the transition to a low-carbon green economy

Specific Capacities

Business skills
Data analysis
Engineering
Entrepreneurship
Environmental and ecosystem management
Environmental knowledge and awareness
Finance skills
ICT skills
Innovation skills
Marketing skills
Project management
Research skills
Sales skills
Science skills
Technological skills
(Gender empowerment skills)

Green Life Skills

Cross-cutting skills that serve both technical, instrumental, and adaptive, transformative ends

Generic Capacities

Adaptability
Collaboration
Collaborative thinking
Communication
Coping with emotions
Coping with uncertainty
Creativity
Critical thinking and reasoning
Decisionmaking
Empathy
Flexibility
Growth mindset
Higher order thinking skills
Interpersonal competence
Leadership
Negotiation
Networking
Open-mindedness
Participatory skills
Problem-solving
Resilience
Strategic thinking
Teamwork

Skills for a Green Transformation

Adaptive skills aimed at transforming unjust social and economic structures

Transformative Capacities

Ability to analyze unequal systems of power
Coalition building
Collective action
Disruptive vs. normative thinking
Environmental stewardship
Future and anticipatory thinking
Integrative thinking
Interdisciplinary and multidisciplinary thinking
Interrelational thinking
Political agency, activism
Reflexivity
Respecting diverse view points
Solidarity
Systems thinking
Trans-cultural, trans-spatial, trans-temporal mindsets
Valuing traditional and indigenous knowledge
Working within complexity

INSTRUMENTAL

TRANSFORMATIVE

FIGURE 3.1 A green skills framework

Source: Christina Kwauk and Olivia Casey, *A New Green Learning Agenda: Approaches to Quality Education for Climate Action* (report, Center for Universal Education, Brookings Institution, Washington, D.C., 2021), https://www.brookings.edu/wp-content/uploads/2021/01/Brookings-Green-Learning-FINAL.pdf.

which types of careers future learners end up in, their jobs will inevitably feel the impacts of climate change. A climate-literate person should understand the principles of Earth's climate system, know how to assess valid and invalid climate evidence, communicate about climate change to broad audiences, and be able to engage in informed decision-making using evidence. Climate science is only one piece of broader systems literacy. Such literacy can be achieved through systems thinking, which that enables us to make sense of dynamic and real-world phenomena and the natural and human-caused factors that affect them.

After witnessing teaching and learning experiences across the Earth Institute and now the Columbia Climate School, we have seen how important these instructional design approaches are in ensuring that systems thinking is incorporated into current and future climate change education efforts. We have also seen that due to the way that climate change through systems thinking has been taught, it lends itself well to promoting change, facilitating co-creation, engaging new audiences, and providing students with opportunities to understand our changing climate through a transdisciplinary lens.

Systems thinking is not only an approach to teaching and learning, it is also an important skill for students as they grapple with the complex challenges that lie at the intersection of climate change challenges and solutions.

The following sections outline evidence-based strategies that may be helpful in bringing systems thinking into the classroom and beyond. They include (but are not limited to) project-based learning, collaborative learning with peers, inquiry-based learning, and opportunities for service learning. We define "evidence-based" here as findings that have been grounded in the field of learning sciences in which different theoretical perspectives and paradigms have been used to build a better understanding of how and where we learn.[5] We have seen these practices integrated into various learning environments and have summarized how they have been used to bring climate change education to learners as well as how they create important learning opportunities in the long run to teach and learn through systems thinking.

PROJECT-BASED LEARNING

Readers may have heard about the importance of project-based learning, but what it means and how it is implemented are important to review. It can be defined as a curriculum framework that helps learners gain a meaningful understanding of content by applying learned knowledge and skills to specific real-world projects.[6]

Project-based learning starts with encouraging learners to demand specific content, and only after they are supported in constructing an understanding of the content can they be better prepared to organize the connections among the content to apply the knowledge.[7] This is particularly important for teaching climate change content. For example, if learners cannot make important personal connections to the content through concrete examples of how climate change may be impacting their communities (and the bigger processes at play that are leading to these direct impacts), we have not opened the doors for them to properly participate in this learning.

With science learning in particular, the authenticity of both the process for doing science and the content itself are crucial. Not only is it important to bring in real-world science content through direct community ties, it is also vital that learners gain knowledge about how scientists are actually doing their work. When we think of science learning, we might picture bench-based activities that have been predetermined, and we can become good scientists if we can replicate those predetermined results. But in reality, the scientific process is much more subjective and often accompanied by questioning, collaborating, and discussing. If learners can be given the opportunity to grasp this process alongside content learning, they can perhaps be better prepared to anticipate and use their gained knowledge.

For project-based learning to be effective, it has to be presented, organized, and sequenced in a way that has a direct impact on learners. With climate change, some foundational content is needed to present information. For example, establishing the fact that our climate is a system is important, and any foundational overview needs to look how the interactions of the Earth's spheres (atmosphere, biosphere, cryosphere, geosphere, and hydrosphere) are also critical. Learners should then have an opportunity to explore additional curriculum that allows them to apply the skills and content gained. This opportunity to apply learned content provides learners with countless possibilities to bring what they know to real-world challenges at different scales and within different communities.

In summary, project-based learning creates an opportunity to engage learners in content as well as offer them opportunities to realize and transfer their knowledge and skills to real-life experiences.

COLLABORATIVE LEARNING

Another critical component is the opportunity for learners to participate in collaborative learning. Climate change content is especially well suited to bringing together diverse groups of learners and engaging them, because the challenges of climate change require interdisciplinary and solutions-oriented thinking that one cannot attempt alone. Learning and problem solving with peers allows for all learners to leverage their collective knowledge and share ideas and skills. For example, some students in a class may be better with physical tasks, whereas others are more theoretical thinkers or planners. Regardless of students' individual strengths, collaborative learning allows them to feel like they are part of a larger collective effort and understand that their role and work is important as part of a broader continuum.

Collaborative learning has also been widely recommended as a method of creating equity because of its ability to create shared roles of participation in classroom settings.[8] Team effort eliminates status differences (e.g., perceived academic ability) in learning environments and interactions and therefore allows for teams to begin work on a more level playing field.[9]

These collaborative opportunities can be facilitated differently in different environments, whether formal classrooms and coursework or informal team learning in other settings. Whatever the environment, collaborative learning requires that learners work as a team on a project in order to achieve specific deliverables. While the content may vary, the work itself will require collaborations over a specific time frame to achieve and ultimately present the results of the project. By allowing students to take on different roles on a team, educators can play to their individual strengths but also ensure that students are working collectively toward a common goal.

This type of learning is particularly well suited for projects that may not necessarily have clear-cut answers but require different considerations to be taken into account. For example, climate vulnerability and impacts is an area of learning that requires a group learning process. Because of the complexity and scope of this topic, collaborations and group settings are more conducive for meaningful discussions and debates, which in turn helps learners develop their ability to communicate and provide evidence to back up claims. This engagement in conversations with others, and not just the content, is also inherently important for learning, as individuals can see the contributions they can make to a complex topic.

Regardless of the different starting points that members of a team may have, the team as a whole is more likely to find its rhythm collectively to meet an end goal as well.[10] Because no one student can be an expert on every topic, the team structure enables everyone to bring something to the table, which in turn gives everyone different opportunities to stay engaged.[11]

INQUIRY-BASED LEARNING

Inquiry-based learning is a teaching strategy that mimics the practices of a professional scientist.[12] Typically, inquiry-based learning has the following phases: orientation, conceptualization, investigation, conclusion, and discussion.[13] Margus Pedaste et al. further divide the conceptualization phase into the subphases questioning and hypothesis generation. The investigation phase has three subphases, exploration or experimentation leading to data interpretation; and the discussion phase is divided into reflection and communication.

This traditional science teaching pedagogy is now also used in social science classes and has been often viewed as a problem-solving technique that includes multiple systematic steps. The application of inquiry learning to climate change education is particularly helpful, as it leads students through important processes of searching for and finding new knowledge, followed by reflection and classroom discussions that can be supplemented with interactions with real-world case studies. Therefore, inquiry-based learning is capable of addressing important competencies in learning about climate change through a systems lens, especially because it facilitates the development of important skills such as creativity, communication, problem-solving, and decision-making, to name just a few. As a way to constantly build and rebuild understanding to further knowledge and experience,[14] inquiry-based learning leads students to explore and expand their understanding on their own.

SERVICE LEARNING

Service learning is a framework that naturally complements climate and earth systems education. It is defined as opportunities and projects that actively engage participants in meaningful and personally relevant service activities.[15] Service learning has a history of success in engaging diverse audiences to participate in civic engagement and can expose learners to diverse values and practices, helping them identify and analyze different views and perspectives.

Much work has documented the positive effects of service learning. At a personal level, service learning improves students' academic outcomes, as they are able to demonstrate a greater level of understanding, problem analysis, critical thinking, and cognitive development.[16] Additionally, students demonstrated that they were more motivated to work harder in a service-learning environment than a traditional class and that their abilities to apply what they learned to the real world were greatly enhanced.[17]

There is also evidence to suggest that when students are engaged in activities that allow them to connect to the learning on a personal level, they not only find these activities more interesting but will also begin to see them as legitimate and in alignment with their cultural and community values.[18]

Service learning can be achieved in numerous ways across different learning environments and levels, and it often involves working with community partners who can offer students learning experiences that are deeply connected to the lived experiences in their own communities. This connection to real-world experiences leads to a level of engagement that is hard to obtain solely through text-based learning that can often feel abstract. Evidence suggests that when used in climate change contexts, service learning improves students' social values and also facilitates involvement in the community.[19] A strong service-learning component fosters the kinds of purposeful learning experiences that makes climate and earth systems science more accessible and available to broader audiences.

In addition to the instructional design components of designing a learning experience, we should also be mindful about the types of learning modalities that allow learners to best give, receive, and store information. In educational settings, those modalities include visual, auditory, tactile, and kinesthetic. Visual learners primarily learn through watching demonstrations that convey a specific concept. Auditory learners learn best by listening in lectures and discussions. Tactile learners prefer to learn through touch and feel (e.g., nature walks). Finally, kinesthetic learners learn best through moving and physical activities (e.g., building a structure).

Climate change and systems science education is made up of complex ideas and processes that have to be conveyed to learners. Therefore, ensuring a mix of learning modalities and incorporating elements of instructional design can help make these ideas more accessible and bring them to life in a classroom in a coherent way that does not overwhelm learners. By incorporating learning science theories to offer different types of opportunities for students to learn new concepts over time, educators can ensure that the teaching and learning that occurs empowers students to learn in new ways, rather than feel anxiety from a doom-and-gloom mentality.

ROLE OF EDUCATORS

None of the aforementioned curricular goals and design strategies are possible without educators, who play important roles in facilitating systems thinking in all aspects of learning. Systems thinking goes hand in hand with interdisciplinary teaching and provides a helpful framework for educators to use to blend natural systems with human, political, cultural, or economic systems. By incorporating systems thinking into classrooms, educators can encourage students to go beyond simplified, black-and-white explanations and move toward creative problem solving outside the usual discipline-based channels.

To facilitate this movement in different learning environments, educators need the time and space to become better prepared for teaching that leads to solving real-life problems in an integrated manner. Place- and problem-based techniques should be favored over traditionally siloed classrooms. This movement toward a holistic and systems approach to teaching doesn't just start with those who are already teaching; it should begin during preservice teacher training. Teachers must have the mindset to learn along with the students and become facilitators in the learning process. Research-based methods, data-based orientation, and scientific evidence are just a few examples of components that need to be incorporated into teacher preparation so that they are equipped with techniques that help instill in students the love of learning through systems thinking.

We need to support teachers and harness their creativity to facilitate student engagement. If we want every student to be aware of the systems connections and impacts of climate change, we need to ensure that every educator is also aware of the multiple dimensions of climate change and its underlying drivers and how to adapt to the impacts. Adapting to a systems thinking approach to teaching climate change means developing in learners across grade levels and subject the knowledge and skills they need to understand climate change. For example, first-grade students who are learning to read can try their skills with stories about animals and the natural environments around them, which provide a perfect connection to exploring environmental issues and climate change in the local community and identifying problems to solve at a higher grade level. With the right support—from materials to coaching—educators can find many connections to learning about the environment and its relationship to climate change in existing structures (e.g., in classroom curricula or after-school clubs, across different subject areas with other educators, or in their own homes). This approach of supporting teachers and students to take action complements curriculum reform efforts aimed at incorporating climate change into classrooms to realize a systems thinking learning agenda.

While there is no one-size-fits-all approach to teaching and delivering a comprehensive earth systems science curriculum, we can apply some of our best practices to focus on student-centered learning. Through the methods outlined in this section, we have seen that inclusive, fair, and accessible learning environments that give all students a chance to learn provide an important foundation for science education, particularly the topic of climate change.

In the next chapters, we will look at local and international examples of how pedagogy and curriculum design has been translated and implemented in teaching practices in formal and nonformal educational environments.

III

EXAMPLES AND CASE STUDIES OF CLIMATE CHANGE EDUCATION IN PRACTICE

4

CLIMATE CHANGE IN FORMAL LEARNING ENVIRONMENTS

In the last chapter, we looked at instructional practices that are particularly useful to consider when thinking about how to best implement a systems approach to climate change education. This is important for many reasons, as these strategies can be applied to a variety of environments for different-aged learners. The strategies proposed are by no means exhaustive, but what follows is an overview of some of the ones that have been documented by practitioners. While those strategies are helpful, it is also important to consider how they might be adopted differently in different learning contexts, schools, and locations.

To engage future generations of learners with the understanding and tools needed to become scientifically literate citizens, we need to ensure that they are engaged with the content across many learning environments at different points in their educational careers. Scientists and educators agree that climate change is one of the most defining issues of our time (and for future generations of learners) and that we are at a defining moment. Our planet is changing; human beings are contributing to these changes; and these observable and measurable changes are happening at unprecedented rates and will ultimately impact the way we live and work in the future.

Education, whether it takes place in classrooms or boardrooms, is a critical tool. Education is needed both to boost resilience and develop the capacity to carry out the enormous social and technological changes that are going to be necessary for human beings to adapt and minimize risks as we face accelerated climate disruption.

We now realize that moving beyond science is needed if we are to achieve transformative and lasting climate change education pedagogy. But to take science beyond just the science department, we must consider what other opportunities students have to engage with the topic, both within and outside school walls. We believe there are three distinct areas and opportunities for learning to occur during a person's educational career: formal, informal, and nonformal education.

We begin with formal learning, which is learning that is done in an organized way, often guided by objectives, skills, and outcomes. For many learners, their first exposure to formal learning environments is through school. Schools have played and will continue to play a significant role in introducing students to and engaging them with systems thinking and climate change, as those settings remain key to helping young people grasp reality amid today's polarized environments and shape the academic, civic, and career paths of learners that will lead them to become resilient and climate literate.

One immediate way to incorporate climate content into the formal educational setting is through integrating climate science and systems thinking into other subject areas besides science. Because climate change is going to affect not just those in the scientific community, we should not be limited to teaching this topic in certain classrooms. Climate change impacts are going to be felt by every human being, and many of these impacts are social in nature. For example, individuals living in populated coastline areas at risk of sea level rise face challenging questions about where their future homes may be. Alternatively, warming oceans are quickly changing migratory patterns of many marine species, which in turn affects the food web as well as entire communities that depend on fishing for their livelihoods. The observed changes may be a scientific issue, but the impacts of the science are going to create adverse socioeconomic impacts in local communities and around the world.

Therefore, we need to move toward broader climate and scientific literacy through interdisciplinary learning and teaching. One of the fundamental and pressing barriers for teaching and communicating climate change is the fact that many people do not believe in its existence despite a clear consensus among the scientific community of many years' standing. Despite the inherent difficulties of teaching climate change, we believe students should learn the scientific basis for climate change because climate literacy provides society with the tools and shared basis for understanding the science and solutions before us.

AN APPROACH TO INTERSECTIONAL CURRICULA

In many curricula across the globe, climate change education is not taught as a separate subject; for example, topics about the environment are now integrated into the sciences and social sciences and not just environmental science courses. The Next Generation Science Standards (NGSS) are built on three dimensions:

- Scientific and engineering practices;
- crosscutting concepts that unify the study of science and engineering through their common application across fields; and

- core ideas in four disciplinary areas: physical sciences, life sciences, earth and space sciences, and engineering, technology, and applications of science.[1]

The NGSS focuses on the natural sciences (physical, life, and earth and space sciences) but also on understanding the human-built world and the importance of recognizing the value of better integrating the teaching and learning of science, engineering, and technology.

UNESCO's Learning Framework on what students should learn about education for sustainable development is a great resource for teachers and school districts, as it provides the learning outcomes aligned to each outcome and assigns them to cognitive, social and emotional learning, and behavioral categories.[2] This framework combines sciences and social sciences for each SDG to provide a set of helpful learning outcomes for students of all ages in formal learning environments.

The goals of climate change education is more than simply to be informative; teaching climate change is different from teaching math equations or Shakespeare works. That's because there is an explicitly prescriptive element: the goal is to produce knowledge but also to inspire changes in attitude and behavior, preparing young people to become tomorrow's (or perhaps today's) climate-conscious citizens, leaders, and problem solvers.

In the next section, we provide examples of how educators from around the world have tackled climate change through systems thinking in their classrooms (and fought off skeptics). But it is important to be aware of what learning looks like in formal environments and why they remain one of the most important sites of education today because in these formal environments, students experience many unique inputs that are not available in other places (at least not all together).

Places of formal education offer a collaborative learning environment that promotes learning and increases a student's awareness about how they as well as others learn. Classrooms also provide students the opportunity to experience social interactions with peers and establish rapport with educators. The socialization aspect of this learning process stays with students for a long time. The classroom also presents students with various scenarios through which they can develop organizational skills in a structured environment (e.g., getting to school on time, finishing homework, and following a schedule). Students learn to that they will be held accountable for their actions through these tasks at a deeper level and learn to be prepared for them.

While these aspects of education in formal settings are not just limited classrooms, they are particularly important when coupled with the presence of teachers, who can modify their instruction quickly based on the types of learners in the classroom (and adapt if they see students having difficulty or finding the topic too easy). When combined, these benefits form the foundation of an important learning environment where students have the time and space to develop as learners.

Despite the advantages of teaching about climate change in a formal classroom setting, many questions remain about how climate change content should be integrated into school curricula. For instance, what should students know about climate change? And how should teachers teach it? While scientists have come to a general agreement about the causes and effects of human-induced climate change, educators continue to struggle with this topic, unsure of whether and how to convey this scientific information to students. Teaching students climate change through systems thinking presents challenges to both teacher and student, not only because the science itself is complex but also because the social forces affecting how participants communicate and respond to the science are complex.

As long as schools continue to remain an important environment where future generations are educated, we need to consider the content that students are learning in these environments that can better prepare them to be responsible climate-literate citizens of the future. Let's now turn to some examples in which climate and systems thinking has successfully made it into classrooms at different grade levels and in different parts of the world.

IRELAND

Primary-school classrooms are not the first place you expect to see climate change content being taught, as arguments are often made for how students are not ready developmentally to absorb this content and that sometimes, it can also be overwhelming if climate change is taught with gloom-and-doom undertones. However, as we'll see in this case, some educators are turning this perception upside down and introducing climate change content in primary classrooms that inspire inquiry, creativity, and cooperation.

In Dublin, the "Creating Futures" module was developed to offer a comprehensive climate change program for upper primary-school students in 2016. The series consists of ten lessons that in reality are mini programs, each likely to be split over several lessons in classrooms.[3] David Selby writes that unlike other climate content, it avoids focusing the learner on just the science behind climate change (e.g., rising greenhouse gases) but is balanced with cross-curricular resources that successfully integrate climate change into other curricular areas.[4]

This is especially important because it presents learners with a systems overview as well as other underlying factors to consider when talking about climate change (e.g., the societal and economic drivers).

As we outlined in chapter 2, a dynamic approach to climate change education needs to complement and be integrated into other subject areas so students can gain the content knowledge and apply it to their understanding of physical and social systems. Climate change is not occurring in isolation of other processes, so we should not be teaching it in isolation of other subjects.

The Creating Futures module employs a range of pedagogical approaches, including small- and large-group discussions, team-based learning, role-play, simulation games, and creative arts and writing.[5] By incorporating these approaches, students never remain in a state of passive learning; quite the opposite. Instead, they are encouraged to ask questions, tackle difficult topics, and debunk prevailing myths, all within a comfortable and structured environment that supports this exploration and sense of adventure.

The series begins with guidance and ten questions for teachers and is presented as an opportunity for them to reflect on their own knowledge and views about climate change. Then the lessons that follow correspond to those questions. By preparing teachers, the modules ensure that they have a solid knowledge base and reassurance to teach a highly politicized issue. Following the teacher section, "Creating Futures" follows a specific sequence to provide a comprehensive overview of climate change, as shown here.[6]

Lesson 1 lays the climate foundations, presenting information students need to know, including the difference between weather and climate.

Lesson 2 covers the exploration of Earth through geologic and human perspectives, then dives into the natural variability versus human causes and effects of climate change.

Lesson 3 begins by presenting evidence that we know climate is changing through climate proxies (e.g., tree rings, ice sheets) and mathematical evidence (for example, plotting the increasing levels of carbon dioxide emissions and global temperature increases).

Lessons 4–5 explore the consequences, impacts, and root causes of climate change. This part brings climate justice into the discussion of climate change and begins to address how those who are most responsible for contributing to greenhouse gas emissions end up feeling the least of the impacts of climate change. A simple activity that allows students to role play conveys important ideas to students about the worst polluters, those who are impacted the most, and fairness and equity challenges.

Lesson 6 looks at biodiversity and climate change through seven activities that emphasize the critical role of nature. Students learn about the significance of biodiversity and have to build a case for why protecting it is important.

Lessons 7–8 begin to focus on changing behaviors (e.g., riding a bike vs. a car) and decisions made today to reverse the course for the future. This turns to a discussion about solutions and highlights solutions-oriented thinking and action.

Lesson 9 looks at the role of leadership and policy-making in climate change and what effective leadership looks like.

Lesson 10 concludes the series on a positive note by asking students to explore their own role in climate change and how they can begin to persuade others (through writing letters to a local or national newspaper) to think differently about their actions around this topic.

UNITED STATES

New York and Gulf Coast—SHOREline

Since its creation in 2013, the SHOREline curriculum has been taught mainly in the Gulf Coast and New York.[7] The curriculum originated from the Gulf Coast Population Impact Project, a four-phase study from the National Center for Disaster Preparedness (NCDP) of the Columbia Climate School to look at the impacts of the Deepwater Horizon oil spill in four gulf states.[8] Years after disasters like the gulf spill, communities continue to struggle with the environmental, economic, and social impacts and often are slow to recover. The SHOREline curriculum is a program for youth that facilitates enhancing their skills, hope, and opportunities as they work toward disaster resilience through community engagement (abbreviations for the "SHORE" part of the name). Ultimately, SHOREline aims to empower youth to build resilience in their own communities after disasters that are sustainable in the long run.

The vision of SHOREline is to develop a network of youth who help themselves, their families, their schools, their communities, and youth in other communities prepare for future threats and/or recover from disaster. Three specific objectives shape the SHOREline curriculum:

1. Develop and refine the skills of SHOREline students and groups to organize and manage themselves, collaborate within their groups, and communicate effectively with community members.
2. Look critically at the world around them and develop insights about the root causes and social forces that may lead to challenges in communities prone to disasters.

3. Develop practical and innovative projects that respond to community problems that can be easily deployed within and outside the home communities in which they were created.

The curriculum content includes establishing and building a SHOREline chapter, developing skills, creating a project, learning about the power of youth and disasters, discovering how to connect chapters, and cultivating professional development. Each section includes instructions for the teacher-sponsor on how to run the various lessons and all readings associated with lessons and activities. Case studies on actual disasters enable students to apply the skills they are developing in disaster management to real-world scenarios. Members of SHOREline chapters gain skills in team management, collaboration, communication, emotional intelligence, decision-making, and problem-solving. The teacher-sponsor guides the chapter and facilitates all SHOREline activities, projects, and curriculum. An approximate timeline is available for the teacher-sponsor, but SHOREline emphasizes that each model varies among chapters and that strict adherence to the timeline provided is not necessary.

Results of the SHOREline curriculum include the creation of a network between chapters, one in New York City and five in the Gulf Coast. The two chapters engaged in a dialogue about the impacts of Hurricane Sandy and Hurricane Katrina in their communities and created a documentary in which students interviewed each other. This output truly embodies the goal of SHOREline for youth helping youth recover from disasters. Further evaluations should be conducted to analyze the short-term and long-term impacts of the chapters on their communities.

New Jersey Climate Education Implementation

New Jersey has integrated climate education into multiple subjects: visual and performing arts, comprehensive health and physical education, science, social studies, world languages, computer science and design thinking, and career readiness, life literacies, and key skills.[9] The state also provides a list of climate education topics to be covered in each grade band: kindergarten to grade two, grades three to five, grades six to eight, and grades nine to twelve. New Jersey has created a hub for climate education resources for teachers such as sample lesson plans, as well as guidance for school boards.[10] Teacher professional development is enabled through a grant from the governor's office.

Each school district in the state follows the standards with regard to integration of climate education. The pedagogy teams for elementary, middle, and high schools develop the scope and sequence of what needs to be taught at which times during the academic year. They also map out the climate education component for each of the standards for each grade. Teachers collaboratively develop their lesson plans and ensure that they are tapping into the standards given by the state.

For instance, the integration of climate science and systems is reflected in New Jersey Student Learning Standards for Science (NJSLS-S). NJSLS-S includes this overarching goal: "All students will possess an understanding of scientific concepts and processes required for personal decision-making, participation in civic life, and preparation for careers in STEM fields (for those that choose)." Desired competencies as a result of this curriculum for students include the following.

- Engage in systems thinking and modeling to explain phenomena and to give a context for the ideas to be learned;

- conduct investigations, solve problems, and engage in discussions;
- discuss open-ended questions that focus on the strength of the evidence used to generate claims;
- read and evaluate multiple sources, including science-related magazine and journal articles and web-based resources to gain knowledge about current and past science problems and solutions and develop well-reasoned claims; and
- communicate ideas through journal articles, reports, posters, and media presentations that explain and argue.[11]

NJSLS-S science standards also specifically describe climate change as the link between human activity and impacts on earth systems. The standards document notes that the "goal is for students to understand climate science as a way to inform decisions that improve quality of life for themselves, their community, and globally and to know how engineering solutions can allow us to mitigate impacts, adapt practices, and build resilient systems."[12] Teachers take these learning outcomes and develop lesson plans using approved existing science materials.

INTERNATIONAL—GREEN SCHOOLS PROJECT

Now turning to the UK, we look at an example of a grassroots effort started by an educator whose felt that the UK government's target of being carbon neutral by 2050 were too slow. Henry Greenwood, who had been teaching math for twelve years, led an environmental program at his school, and then decided to start the Green Schools Project in 2015. The project's mission to enable young people to fulfill their potential by providing UK schools with resources and support to aid them in creating environmental projects, building critical skills, inspiring students, and encouraging students and their communities to live in a more sustainable way.[13]

The Green Schools Project uses the Zero Carbon Schools framework, which empowers students to drive change and improve environmental awareness in their school, community, and world through reducing carbon emissions.[14] More than twenty schools are actively running green clubs modeled after the Kingsmead School's program, and the 2021–2022 school year will see an expansion to London schools.[15] The goal is that all schools involved with the Green Schools Project be carbon neutral and for students, teachers, and the community to be invested in connecting with nature and fighting the climate crisis.

Using Kingsmead School as a case study, remarkable improvements have been reported over three years. Before the Green Schools Project began at Kingsmead, the biggest issues highlighted were the lights left on in empty classrooms, windows left open during winter when the heat was on, recycling bins rarely emptied, litter, and no student engagement in environmental issues besides the topics studied in geography and science.

ONLINE RESOURCES TO SUPPORT CLASSROOM TEACHING AND LEARNING

An incredible array of online resources is now available to support the teaching and learning of climate change in formal educational environments. The following examples are just a few highlights that draw on online resources that have become popular and useful for high school science educators in the United States. These resources are an initial set of modules that have been developed as specific units in science classrooms. The focus of most of these resources is on ensuring that the content matches grade-level expectations and standards that formal educators have to grapple with at the high school level.

A compelling introduction to climate change that emphasizes the data and evidence that has been gathered to prove our climate is changing is a primer from the U.S. Forest Service Climate Change Resource Center that introduces the topic well. The Climate Science Primer is a web-based work that includes data-rich graphs and research that addresses the correlation between temperature and atmospheric CO_2 witnessed over the past 450,000 years, the rise in sea level since 1870, and the decline in the Antarctic ice mass.[17] The primer also explains the mechanisms behind global warming (e.g., the greenhouse effect) and the human influence on the greenhouse effect, along with natural climate cycles and how what we are witnessing now are unprecedented.[18]

Another relevant resource has been developed by the National Aeronautics and Space Administration's (NASA) Global Climate Change: Vital Signs of the Planet website.[19] The site provides graphics and multimedia resources to help students visualize climate data, and they can access short videos featuring satellite imagery and images from space, as well as infographics on the rise in global temperatures and sea level.[20] The real-time nature of the data on this site is attractive, and it presents teachers with various learning modalities and tools with which to engage students that are not just text-based.

Although a multitude of resources exist on how and why climate change is occurring, relatively few provide details about the impacts of climate change on plants and animals.[21] *Ecological Impacts of Climate Change* from the National Research Council fills that gap completely for the United States. This resource presents examples of impacts on specific plant and animal species in each of seven U.S. regions: the Pacific coastline, Alaska and the Arctic, the western mountains, the southwestern deserts, and the central, southeast, and northeast regions.[22] This consensus study report features engaging photographs and other visuals and connects important concepts on climate change and ecology.[23]

Different forms of media are important to consider in student learning, which is why short videos are such a great resource. The U.S. National Park Service has produced a series of movies (3:05–8:58 minutes) on the consequences of climate change for plants and animals in national parks across the United States.[24] Topics in these videos include ocean acidification processes and how that impacts marine ecosystems worldwide.

It can be very time consuming to look for resources in an ad hoc manner across a number of sites, which is where searchable databases can be useful.[25] The Climate and Energy Educational Resources (CLEAN) Collection[26] is one of the most extensive collections in existence, featuring hundreds of resources and activities arranged in categories (e.g., short demos, experiments, visualizations, and videos). What is unique about the CLEAN database is that qualified scientists and educators from organizations that include the National Oceanic and Atmospheric Administration, the Science Education Resource Center at Carleton College, and the Cooperative Institute for Research in Environmental Sciences continuously update the resources.[27] Rigorous review processes, similar to those for scientific publications, are equally important in ensuring that the science content provided in educational resources is accurate and aligned with important frameworks, like the United Nations Sustainable Development Goals and the Next Generation Science Standards.

Educators are in a unique position to tackle the challenge of introducing the topic of climate change by addressing misconceptions and deliberate misinformation perpetrated about it,[28] and conveying that knowledge does not have to take place in a science classroom. Engaging people in discussions is worthwhile, because if you're not speaking to the "other side," there is very little chance of addressing those misconceptions.

ClimeTime is another useful resource, facilitated by the U.S. Office of the Superintendent of Public Instruction (OSPI). This initiative is run in collaboration with the University of Washington Institute for Science + Math Education

> through a Washington State legislative proviso originally requested by Governor Jay Inslee of an annual $4 million investment that began in 2018–19 and continued as a $3 million investment in 2019–20. OSPI manages the network, and the grant funding flows through all nine Educational Service Districts (ESDs) in Washington and six community-based organizations (CBOs). The ESDs and CBOs have launched programs for science teacher training, linking the Next Generation Science Standards (NGSS) and climate science. In addition to teacher professional development, the project supports the 15 grantees to develop instructional materials, design related assessment tasks and evaluation strategies and facilitate student events.[29]

Skeptical Science is a great resource for the classroom, providing peer-reviewed content organized around claims that are most frequently made by climate change skeptics.[30] The site's resources use a variety of mixed media (e.g., text, videos, data visualization) to communicate and refute the most commonly used arguments by skeptics. In addition, almost one hundred short lectures and forty full-length interviews with climate scientists are specifically designed for teachers to help them introduce the topic by first addressing deliberate misinformation that has been perpetrated about climate change.[31]

TROP ICSU, or the Trans-disciplinary Research Oriented Pedagogy for Improving Climate Studies and Understanding, is a global project funded by the International Council of Science.[32] The project is led by the International Union of Biological Sciences and coled by the International Union for Quaternary Research INQUA. TROP ICSU collates climate studies across the curriculum of science, mathematics, social sciences and humanities. These teaching resources are locally rooted in their context but globally relevant with regard to science.

Green Ninja[33] is a phenomena-based science program for grades six through eight. The program engages students by providing them with opportunities to use science and engineering to solve real-world environmental problems. The program is thoughtfully packaged and flexible, making it easy for teachers and students to access content through a digital platform and printed materials. The curricular resources are developed by Eugene Cordero, a climate scientist at NASA.

Some resources are focused more on sustainable development, like SDGs Today.[34] SDGs Today is a part of the United Nations Sustainable Development Solutions Network (SDSN), in partnership with ESRI and the National Geographic Society. This resource is a global hub for real-time SDG data. The mission involves advancing the production and use of real-time and geo-referenced data for the SDGs. In partnership with ESRI, the initiative has been able to provide lesson plans (learning paths) advancing understanding the SDGs as well as involving students and teachers to use their voices locally through their story maps.

More local stories that are student driven are narrated at Design for Change.[35] This resource offers initial training for teachers and students using the framework of "Feel, Imagine, Do, Share" (FIDS), with empathy as the driving force to turn students into local change agents.

As we review these online resources, it is important to keep in mind that they need to be mapped to the state's or nation's curricular standards and become a part of the lesson plan to be taught through each school's formal curriculum. In many cases, standalone videos on YouTube that discuss important climate topics may not come to the notice of teachers while they are developing their lesson plans. Similarly, while students doing their online searches may not spot such videos. With the plethora of resources available, the chances of their discovery by teachers depend on the standard mapping and lesson plan integration.

Classrooms and other sites of formal learning provide ideal settings for students to elaborate on existing beliefs, change perspectives and approaches to beliefs, and take up new beliefs. And these types of opportunities are available across all classes, not just one focused on science. The tensions behind motivating students to begin to discuss and address climate change and its distressing realities are very real, and we have learned that the gloom-and-doom approach is not necessarily the best one. Instead, posing questions to students and allowing them to pose questions about the obstacles they see and the concerns they feel foster a more effective learning environment.[36] Experts note that current classroom teaching is about unsustainability, not about sustainability or solutions for a resilient world.[37]

Beyond addressing climate science skepticism, climate change education should invite new experiences and thinking. The sociologist Jack Mezirow says that certain educational occasions can pose disorienting dilemmas, instances in which initial feelings of discomfort are unveiled that provide the raw material and foundation of productive reflection and reorientation. From these opportunities and unsettling emotions can grow transformative learning opportunities for students, in which they can question, change taken-for-granted assumptions, and forge new beliefs.[38]

The scientific community can stand to learn a few things from the social science community in this realm. Instead of just leaving students with the irrefutable evidence that our planet is changing (and not in a good way), how can we spur them to new forms of agency and action that they can try out in the safety of a classroom space and then take out into the world? Students want to engage, and their anecdotes in nonscience classes are often indicative of the intellectual curiosity and fortitude they possess, especially when given the chance.[39] Taking a transdisciplinary approach to the topic will help to provide a holistic and in-depth treatment of it. However, the reality is that subjects are taught in an insular manner; students gain siloed knowledge with little or no applicability in the real world.

As Siperstein outlines,[40] the same aspects that make climate change difficult to teach at first glance—its multisystem complexity, its political and ideological baggage, and its plethora of depressing feelings—are precisely what also makes it relevant and meaningful to a wide range of students. To truly maximize students' potential for intellectual and emotional growth, interdisciplinary learning environments need to be considered holistically, including the intellectual, social, emotional, and physical space in which the learning occurs.[41] Next, we highlight an example of curricular activities that have ample scaffolding to encourage students to develop agency through interdisciplinary settings.

Climate Action Kits

TODAY: Growing Justice

Level
Primary

Big Idea:

Pollution is anything that makes our earth unhealthy and dirty – like smoke, trash, and other stinky stuff. Pollution can hurt our air, water, and the soil where we grow our food! Some communities are hurt more by pollution than others, and this is called environmental injustice.

Scientists are working hard to find better and better ways to help clean up our pollution, and one beautiful, eco-friendly way to clean soil is by planting...sunflowers!

Taking care of plants at home can also help us do our work well, and makes us feel good!

You will need:

- Scissors, Glue, Pencil ✓
- Peat pot & soil ✓
- Sunflower Seeds ✓

ICE BREAKER: WHAT IS ENVIRONMENTAL JUSTICE?

Now that we know sunflowers are great at cleaning polluted soil, we must ask ourselves how these toxins got into the soil in the first place? Usually, the answer is: through humans! When big companies and businesses are producing the things we buy, many times they discard toxins in ways that are bad for the environment and can hurt plants, animals, and other humans.

The call for environmental justice comes to try to fix this problem so that everyone can live their healthiest lives! Justice means fair treatment for everyone, and environmental justice happens when every person acts together to make decisions about laws regarding the environment. The environment includes forests and lakes, but also houses, schools, neighborhoods, and cities. If everyone works together and listens to each other, it is less likely that some people will do things that hurt other people – like discarding toxins in water and soil!

SDG Students Program

Center for Sustainable Development
EARTH INSTITUTE | COLUMBIA UNIVERSITY

ECO AMBASSADOR
ENVIRONMENT IN ACTION

FIGURE 4.1 Sample lesson plan on justice

Source: Tara Stafford Ocansey, Eco-ambassadors and Mission 4.7. Growing Justice. Eco-ambassador and Mission 4.7 internal working documents (unpublished).

"The Growing Justice" lesson plan was developed for lower elementary grades under the leadership of the Center for Sustainable Development at the Earth Institute. The main intention behind it was to provide a sample of an experiential learning pedagogy with an interdisciplinary approach.

This lesson plan for primary grades introduces a powerful concept of environmental justice (figure 4.1). It has been carefully crafted using scientific knowledge on pollution and how it affects the air, water, soil, and food. To make the narration positive, the lesson includes planting sunflowers as a solution for pollution. The concept of environmental justice is introduced as an icebreaker to discuss unfair treatment of certain subsections of the population. Awareness about where toxins are dumped and how pollution is caused is presented through an environmental justice lens.

The lesson plans provide activities using simple kits introducing ideas of social-emotional learning, STEM, and civic engagement. Biological aspects of growing a sunflower and linking it to reduction in pollution provides knowledge of science concepts as well as an action that can be carried out easily in yards, gardens, or community spaces. The students also get to improve their English vocabulary as well as write a journal on the growth of the flower using measurements (math). This lesson plan is also mapped to the New Jersey Student Learning Standards, which closely aligns to the Next Generation Science Standards. Thus, mapping to the standards ensures that the lesson plan is integrated into what the students should be learning at each grade level.

This chapter provided a glimpse of programming and lesson plans that introduce many elements of a systems thinking approach in climate change education around the world. This is not an exhaustive list but an introductory overview of how systems thinking is being incorporated into climate change education in a transdisciplinary way across subject areas and classrooms. These examples also demonstrate, in action, the many instructional practices outlined in the previous chapter (e.g., learning and teaching through experiential and project-based learning, collaborative learning). Our intention is to get educators to start thinking about the ways in which current classes and curricula can be adapted (and not completely wiped out) to incorporate important content. Our aim is not to suggest revamping everything that is happening in a classroom but to brainstorm and offer examples of how systems thinking around climate change can be integrated into existing curriculum.

An important point to note here is that the best lesson plans always come from the teacher, who knows the students and community and can reflect them in the lessons. This chapter is only a starting point toward providing initial ideas and examples, and our hope is that through this important work, all classrooms can begin to include elements of earth systems science in lessons.

5

COMMUNITY-BASED (INFORMAL) EDUCATION

The previous chapter provided an overview of important examples from formal learning environments that demonstrated how educators were working in a bilateral way (both grassroots and top-down) to bring a systems-based approach to climate change into classrooms and schools around the world. While the school and classroom setting remain one of the most important educational environments, we also need to consider the importance of the learning that happens outside the classroom. In this chapter, we will turn to community-based and/or informal learning, which is education that goes beyond the formal learning environment and can occur in daily life experiences through peer groups, family, media, or any other influence in the learner's surroundings. These learning opportunities are less organized and structured in terms of the outcomes you'd expect from a formal educational environment. Many of these experiences are referred to as "learning by experience." The idea is that simply by existing, the individual is constantly exposed to learning situations at work, at home, or during leisure time, for instance.[1]

Many important stakeholders are involved in informal and community-based education, including but not limited to families, libraries, and museums. The learning in these settings is mostly experienced and lacks the formalities of classroom schooling, such as following a schedule or taking an exam after learning a complex curriculum. The key word here is "experience." One can expect to experience informal learning by going about their daily life and how they may be influenced by who they interact with and what they see in their communities. Therefore, we believe that much of this type of learning occurs in our communities, whether from our parents and elders or from museum exhibits. The commonalities of informal learning efforts include a boundary-free environment where one can choose how and what knowledge one receives from experiences and encounters faced in everyday life.

THEORETICAL FOUNDATIONS

The theoretical underpinning for community-based informal learning efforts is linked to the concept of "funds of knowledge."[2] The concept of funds of knowledge is defined by Moll and colleagues as "these historically accumulated and culturally developed bodies of knowledge and skills essential for household or individual functioning and well-being."[3] These funds of knowledge are the result of people's lived experiences, including their social interactions, cultural and religious backgrounds and practices, their participation in multiple labor markets, and their varied language-related activities.

Funds of identity, related to funds of knowledge, are strongly affected by cultural factors such as sociodemographic conditions, social institutions, artifacts, significant others, practices, and activities. To understand individual funds of identity, we first need to understand the traditions, beliefs, knowledge, and ideas that contribute to them. As such, lived experience results from any transaction between people and the world, emphasizing the subjective significance of the situation to the person. It is the subjective side of culture, which mediates and organizes behavior.[4] Funds of knowledge become funds of identity when people actively internalize family and community resources to make meaning and describe themselves.[5] It can be argued that identities, created and recreated in interactions between people in every context, are lived experiences of self.

In both individual and collective settings, education is an integral part of the lived experience. In a community setting, "education" is defined by the learning experience in which influencing and informing factors come into play, regardless of where it takes place. Hence, such education is a learning experience in households, between peers, and in the community, not confined to formal school systems and classrooms. Researchers recommend maintaining a level of cultural integrity, which includes viewing education not simply as a process of passing information to parents without regard to their cultural realities but, rather, as an interactive process of identity and community development that respects the culture and knowledge of the family.[6] That interaction can occur through daily educational experiences with the households along with the interactions with extended family members.

TRANSFERRING FUNDS OF KNOWLEDGE AND FUNDS OF IDENTITY TO THE LEARNING EXPERIENCES

Funds of knowledge and the process of transmission and learning can facilitate a powerful and culturally relevant way to tap into communities' resources in the classroom.[7] Furthermore, when funds of knowledge are incorporated into the learning experience, they enable learners to gain knowledge from multiple spheres of activity with family relationships, social worlds, and community resources rather than a single-stranded relationship between the student and the teacher.[8] Thus, a process that acknowledges and incorporates the resources, interests, and values of families and that uses the funds of knowledge framework can create a meaningful learning environment.

Pedagogical approaches grounded in learners' cultural backgrounds and everyday knowledge can make a difference in learning[9] and foster greater understanding between and among teachers, students, and families. School-based education has the potential to build a robust educational community environment that links students' cultural knowledge and experiences with more traditional learning spaces and contents. The result would be a bridging of learners' worlds with fundamental environmental and social issues can support them in building deeper scientific, social, and civic understanding and fostering their awareness of their agency. Therefore, learning should be encouraged not only through a cognitive lens but also within a community. Environmental research will foster empirically grounded discussion about promoting learners' empowerment and engagement as intertwined deep and meaningful learning elements.

Understanding the deep-rooted factors that shape an individual's lived experience will lend itself to the type of learning experience and education that can speak to individuals in meaningful and engaging ways to promote engagement and behavior change.

Indigenous Knowledges

Not only does indigenous knowledge need to be integrated into the school curriculum, but also its presence in the community needs to be recognized. Research suggests that indigenous knowledge helps connect humans to nature.[10] Through this interaction, views, aspects, and practices from indigenous communities are integrated into science teaching and learning. Indigenous knowledge provides the content enabling one to connect to one's immediate environment and provides authentic science learning. Robby Zidny et al. have argued that indigenous knowledge helps create more balanced and holistic worldviews and intercultural understanding of sustainability.[11]

Indigenous knowledge is based on a relationship of reciprocity and caretaking. These principles help encourage a sense of unity and interconnectedness and a desire give back to nature.[12] Many researchers have used indigenous knowledge to form the core principles behind sustainable natural resource management practices and rural livelihood.[13]

Over the years, indigenous knowledge was seen as inferior to scientific research.[14] However, researchers suggest that multidisciplinary and interdisciplinary approaches, not just scientific inquiry, should be used as tools for solving environmental sustainability problems.[15] Indigenous knowledge is a vital source of content that can provide guidance toward long-term solutions through ecological responsibility, and such knowledge has been underutilized in informing informal, nonformal, and formal education systems.

Challenges

Community-based and informal education faces a fair set of challenges. As a result of mandated standards and testing and predetermined competencies, there is little room for teachers and schools to make essential connections between formal and informal community learning. In the case of teaching climate change and earth systems science, this creates a gap. Because community needs and experiences and classroom learning do not always intersect, learners cannot see the connections between the systems within their communities. This disconnect between the schools and communities needs to be addressed through different avenues, and numerous organizations and efforts are under way.

The following are just a few examples of how organizations are strengthening community-based informal education through efforts grounded in the funds of knowledge pillars of experience, awareness, and engagement.

COMMUNITY-BASED AND INFORMAL EDUCATION PROGRAMS

Sin Planeta B ("No Planet B")

The organization Sin Planeta B conducts research in Mexico to see how educating children about climate change impacts their parents' lifestyles and behaviors. Through observation, interviews, and questionnaires, it monitors the extent to which this education can effectively create a more sustainable future.[16] Places without these programs can still take substantive action toward a green agenda and use the power of climate education through developing a coalition for action, supporting teacher and student creativity, and capturing learning to advance impact.

Children have considerable influence on their parents, especially on more controversial topics that their political ideologies ordinarily prevent them from engaging with information.[17] The work of Sin Planeta B is critical because in Latin America, families are more involved in their children's education and development compared with families in other continents.[18] Sin Planeta B provides a unique opportunity to target children and adults in climate change education and promote sustainable behaviors. These children will have a more significant influence on their families than children on other continents. Parents have reported to Sin Planeta B that their children invited them (or nearly forced them) to participate in environmental activities and discussions in a collective way. As a result, children feel more empowered to be agents of climate education to promote changes in their parents' behaviors.[19]

The research from Sin Planeta B is still in its early stages. Still, the evidence suggests that the power of educating children on climate change will extend to their families and have a significant impact on promoting sustainable lifestyles through this intergenerational learning. Moreover, as the climate crisis becomes a more dire issue and the need to address it intensifies in the next few years, a national program on climate change education would only expand the beneficial impacts already being seen across families. Children are the next generation, and empowering them through climate change education is essential for continuing to promote new ways of thinking and attitudes toward global climate action.

Eco-Ambassadors Program

Originating from the Center for Sustainable Development at Columbia University's Earth Institute, the Eco-Ambassadors effort is situated in Millburn, New Jersey, where community-driven environmental actions are taking place and growing. The team behind this effort has been involved in literacy initiatives internationally and has contemplated using environmental material in their literacy approaches. The program emphasizes getting the community and students interested in scientific programming.

Workshops began in summer 2019, in which students participated in various environmental projects. Weekly activities, meetings with scientific experts, and sharing of personal projects were just a few of the critical elements of this effort. Students and families engaged in science through a variety of research- and action-oriented projects focused on identifying and responding to community challenges. By providing students with the spaces to share their work (e.g., the International Conference for Sustainable Development during the United Nations General Assembly Week), online and social media platforms for learners to engage in the science alongside their family members, and opportunities to connect to real-world issues, this program demonstrates the importance of informal learning through natural experiences within communities.

As a result of these community-organized efforts, the Millburn township passed a single-use plastics ordinance in June 2020, which prohibits single-use plastic carryout bags and polystyrene containers at retail establishments and reduces the use of single-use straws within the township. Led by the town's Environmental Commission, the ordinance passed after much vocal organizing by key community members, including parents and middle and high school students. In addition, this policy has encouraged many other environmental stewardship actions, including eco-ambassadors conducting reusable cloth bag marches and talking to the vendors to reinforce the plastic ordinance. Eco-ambassadors organized campaigns in town to avoid single-use plastic, and students organized reusable cloth bag sales before book sales events at schools. Community-led efforts on environmental topics were noticed by the Millburn Education Foundation, which began hosting an annual STEM (science, technology, engineering, and math) environmental challenge in 2020 to present.

EI LIVE K12

Another successful public outreach effort at the Earth Institute is the EI LIVE K12 (now known as Climate Live K12) series. Named for the Earth Institute and implemented during April 2020 when the COVID-19 pandemic had reached New York City's doorsteps, the series began as a way to quickly supplement science learning for K–12 students and educators who were interacting remotely.

With sixty-plus sessions that have reached more than eighteen thousand individuals, the series is here to stay. The series brings the science behind the climate crisis and sustainability virtually to audiences nationally and internationally. The K–12 channel features experts from around the institute and Climate School presenting relevant sustainability content in live sessions of forty-five to sixty minutes for students, parents, and educators.

Each episode that has aired follows a specific format. Content experts typically begin with an introduction or overview of the topic, share recent findings, and, if applicable, conduct a relevant hands-on activity. Once the content part is completed, we open it up for audience questions, which is carefully facilitated to ensure that they have the time and space to have their questions addressed by our scientific experts. Following each session, recordings are shared with all viewers, along with additional reading material and activities where appropriate.

The programming offered through Climate Live K12 is often a balance between traditional lectures, hands-on activities, and demonstrations. The limited enrollment allows for an intimate environment that gives viewers the time to process the information and ask questions regardless of the program's format.

These informal education programs help in reaching out to general audiences and educate communities about climate topics. It is also important to recognize that communities need some form of institutional support and social capital to make climate education possible. That may come from existing nonprofit organizations, public libraries, youth leaders and change makers, and professional experts (e.g., agriculture extension workers, forest guards, and agriculturalists) who reside in the communities who have vast experience in conservation and other climate topics. The next section lists examples of community-based resources that can be tapped to promote climate literacy.

BHOPAL'S SOLUTION TO THE PLASTICS PROBLEM

The nonprofit Mahashakti Shakti Seva Kendra was recently featured in UNESCO's trash hack mission.[20] A group of local women activists formed the organization in 1994 in the wake of the Bhopal gas tragedy. The main task is to provide employment to women and make them economically self-reliant. The organization has received many grants from the government to work with natural dyes and cloth printing, as well as stitching and tailoring. Along with the organization's mission of "No more chemicals," the group also began to focus on making bags out of discarded waste, primarily because the Bhopal gas tragedy has taught them to be more environmentally conscious but it was also cheaper for them to make bags out of what was considered waste materials.

The fifty women who work at the center collaborated with the resident welfare associations (RWA; apartment complexes usually have an RWA to examine the building's needs) to collect old clothes and transform them into grocery bags for the residents. They also collaborated with the town's municipal corporation to make garbage bags for the entire city from vinyl posters that are left on roads after an event. They learned about plastic waste and transformed their homes using minimal plastic projects, starting with minimizing plastic kitchen utensils.

The informal environmental education gathered from their efforts led one community to act together. The women at the center also came up with the idea of eco-friendly shops with upcycled products in various parts of the city, as well as hosting street theaters, artists, and thought leaders to talk to city residents about the recycle-reduce-reuse concept.

The community volunteers who led the efforts received no formal training about plastic pollution. Only about half the women at Mahashakti have graduated primary schooling, and around 60 percent can read Hindi at a basic level. They are strong supporters of plastic control in the city and have created a source of income through their green skills.

THE ROLE OF LIBRARIES

The disengagement in STEM among women, Latinos, and African Americans is higher in the United States than in other countries, and the level of pursuit of secondary degrees in those fields is also low for those communities.[21] The several barriers to historically underrepresented communities include cost, lack of a safe way to get to and from after-school programs, and lack of support in their interest in STEM. The report *STEM Equity in Informal Learning Settings* examines public libraries' role in increasing STEM equity and access for historically underrepresented K–12 students. It notes that with fewer individuals pursuing secondary and graduate degrees in STEM-related fields and a shortage of qualified professionals, informal education could solve this deficiency by providing education and programming at different levels and platforms, increasing accessibility to STEM education.

Informal STEM education is not a replacement for but an addition to formal STEM education. Libraries have long been sites of informal education, with their constant supply of resources that provide a foundation for STEM education, such as internet access, computers, books, and knowledgeable staff. In addition, libraries are community institutions, a well-known place to gain access to books and magazines and to socialize through community events and after-school programs.[22] Suggestions for improving engagement with these communities include mentoring, tutoring, career counseling and awareness, learning centers, workshops, and seminars—all of which can be offered at a low or no cost for students.

This report used a literature review to shape the following recommendations for public libraries to incorporate into their programming:

1. collaborate with STEM stakeholders;
2. form partnerships with organizations that serve youth;
3. target historically underrepresented K–12 youth;
4. make STEM programs accessible and equitable to all youth;
5. develop strong, lasting, caring adult–youth relationships;
6. provide training and professional development opportunities to librarians;
7. evaluate STEM programs and monitor and track outcomes; and
8. share results with stakeholders.

Many recommendations have been made and much research conducted about how to engage and maintain students' engagement in STEM fields and the reasons STEM engagement falls off as students grow older. However, there is always a need for additional research into what engages students. *STEM Equity in Informal Learning Settings* states that public libraries are the next area of research in this sector, and evaluating the effectiveness of STEM activities and services is the place to begin.[23] Though the recommendations are targeted at librarians who want to enhance the STEM education and programming in library settings, policy makers, researchers, and practitioners are able to use these tools to create best practices and design and implement programs in their communities.

Several library organizations engage in research, provide resources, and support the library community in STEM efforts through conferences and blogs. The American Library Association (ALA),[24] the Young Adult Library Services Association, and the Association for Library Service to Children lead the way in supporting informal educational opportunities. The latter two associations provide toolkits designed for professionals to incorporate in various learning environments and a variety of learning experiences and archived webinars for "How to Put the Library in STEM."[25] These two organizations provide resources but are not explicitly involved in providing funds or aiding in developing specific libraries' programs. In contrast, ALA is directly funding informal educational opportunities.

ALA hosts a number of free informal opportunities ranging from kindergarten to college level in rural communities. Their STEAM (science, technology, engineering, arts, and math) program is free, virtual, and recorded and focuses on bridging the gender and race gap in STEAM education. In addition, ALA offers five webinars with topics like "Closing the Gender Gap:

Developing Gender Equitable STEAM Programs" and "Culturally Responsive STEAM Programming: Engaging Latinx Communities in Rural Areas." The purpose of the STEAM series is to collaborate and cocreate gender-equitable exploration pathways with community leaders and youth, grow the ability of community collaborators to initiate and sustain STEAM opportunities, and actively engage youth and their families to encourage them to participate in STEAM activities.[26] More than twelve libraries received grants of $15,000 for four years beginning in 2020 to enable them to participate in the STEAM program. After that period, the STEAM learning materials will be assessed and reevaluated.

This report also cites four promising STEM programs currently implemented in public libraries around the country. STAR Net Libraries; YouMedia Lab Learning Network; Learn, Explore, and Play (LEAP); and Explore Library Program established informal STEM education activities for youth.[27] STAR Net is a national informal education and outreach program designed for library settings. It provides resources, activities, professional development, and librarian training to build the capacity for public libraries to provide STEM education. Evaluations showed that six months after implementation of STAR Net programs, a positive impact was seen in an increase in interest and engagement of library patrons in the STEM topics, indicating the potential for long-term positive effects of the program.

YouMedia Lab Learning Network provides STEM-related experiences for middle school and high schools students through technology and digital media for libraries and museums. It aims to expose students to new interests, develop their skills and expertise, and connect these interests to future career opportunities. In addition, the YouMedia Lab Learning Network focuses on mentorship and providing tools for further success in personal and academic life.

Learn, Explore, and Play (LEAP) aims to connect STEM subjects to everyday life for students age eight to thirteen through collaboration with public school teachers and STEM professionals to create STEM kits that students can check out of the library. A more independent activity, LEAP incorporates inquiry-based learning while allowing students to choose their kits.

Finally, the Explore Library Program, created by the Lunar and Planetary Institute, offers more than one hundred free space science activities and educational support resources for children and preteens. In addition, the program provides library professionals with materials, content knowledge, and skills to facilitate STEM training for youth. Nine hundred-plus librarians and educators across thirty-five states are participating in the program. Additional research is needed to determine the impacts on youth, but initial results indicate that educators and librarians have seen significantly higher levels of earth and space science knowledge.

WORKING WITH LOCAL CHANGE AGENTS

The International Research Institute on Climate and Society (IRI) at the Earth Institute has a long history of working across the science, policy, and decision-making aspects of climate science. For nearly twenty-five years, IRI has been one of the leaders in generating climate forecasts, training national meteorological and hydrological services personnel, and supporting the development of systems for climate services tailored to specific partner decisions. In the context of disaster preparedness, with the new subseasonal-to-seasonal (S2S) forecast system, IRI has proposed a concept called "Ready-Set-Go!" for using forecasts issued with two-week to two-to-three months lead times to help make different sectors make important decisions about risk reduction. For this activity, IRI is closely collaborating with the Red Cross Red Crescent Climate Centre, some of whose staff have cofunded positions and secondments.

This concept could be used in flood preparedness, for example, whereby large-scale climate indexes, catchment wetness, and observed rainfall events at the seasonal lead time can encourage humanitarians to update their contingency plans and early warning systems in the "Ready" phase. Then, semimonthly, S2S forecasts enter the "Set" phase, where they are used to preposition materials, alert volunteers, and warn communities and decision-makers about an increased risk of flooding. Numerical weather prediction weather forecasts and warnings would fuel the "Go!" phase, as they are used to activate volunteers, distribute instructions to communities, and evacuate areas if needed. In addition to flooding, these forecasts can also be used for other sectors, such as agriculture.

While high-quality climate science is vital for the potential success of such a system, policy development around sustainable, accountable government systems, and community stakeholders are equally important. With the strengths of these examples come opportunities for improvement. For instance, we have learned that there are significant challenges to maintaining these systems if and when the technical assistance of the climate science advisory team decreases its involvement. A structured approach is needed for developing, deploying, and assessing educational mechanisms for subnational-level stakeholders to drive this type of work. However, these partnerships across sectors, climate and social scientists, and communities are critically important to advance community education efforts around climate change.

WORKING WITH THE DISTRICT AND STATE GOVERNMENTS

Earth Institute partners with many organizations' grassroots initiatives on research. It also works with various nonprofits and local government agencies on multiple research projects. The government department that leads the interventions can be at the district or village level. Dialogues about policy around sustainable development occur at the state and national levels based on the action research.

Working with a government is advantageous for many reasons. First, the EI-led studies can be integrated into the government's systems rather than stand-alone projects that are not meant to last. Second, since most of the institute's work is multidisciplinary, involving the departments within the government that are best suited to work together to solve a sustainable development problem. Finally, most of the research aims to build capacity with a big push toward participatory action research where the stakeholders (government and the community members) are involved in the research process as active members. Working with various levels of the governmental hierarchy also helps update skills as well as achieve the action research component. During discussions at all levels, many departments are usually involved, including water, education, health, forestry, rural development, and data and statistics.

Since sustainability is intersectoral, many of the Earth Institute's centers also have a multidisciplinary team that works on field-based research. Often, these interventions bring the community together and involve some elements of community ownership. The collaboration of research and practice helps in capacity building at the institutional level, but it also helps position the community at the center with the planned interventions. The shared-solar project is one such example.[28] This concept at the Quadracci Sustainable Engineering lab looks at low-cost, incremental power infrastructure for providing affordable, grid-like electrical services by combining renewable energy with smart metering and storage management. In this way, reliable and verifiable services can be provided to off-grid communities in remote and rural areas. The system uses a micro-grid network powered by solar panels to deliver power to consumers, and each household uses a unique prepaid metering system. The basic idea is to make electricity affordable for all. The project was implemented in selected rural communities in Mali, Tanzania, and Uganda. Many such initiatives help communities become empowered using community-based skills and affordable technology that can be scaled up.

Another case study is based in Nigeria. CSD and CIESIN at the Earth Institute collaborated on the Nigeria-based GRID3 project on spatial planning for education decision-making.[29] The researchers conducted workshops with local- and national-level education planners in Nigeria using geospatial maps. The maps aided in district and statewide planning on SDG education. This basic function of getting all schools on a geospatial map can make education planning data oriented. Locations of schools were mapped over school-going population age groups, which measured how many schools were overcrowded. The research team discussed the geospatial education planning scenarios with education planners at the national and state levels to find the more nuanced stories. Ministry of Education staff discussed the feasibility of using data-referenced planning for education at the national and subnational levels in Nigeria.

YOUTH ACTIVISM

Community organizing requires specific skills to make people interested in your story, gather momentum, and bring change through a particular action, demonstration, or strike. Though these community-organizing skills are not taught in schools, the youth in many countries have demonstrated the impact of large-scale mobilization, such as Greta Thunberg's #Fridaysforfuture strikes. Students learn these skills on their own and pass them on to others who want to be active in this space. Some students are organizing fashion shows around the idea of converting trash into fashion.[30] In the Climate Education Youth Summit for New Jersey and New York, student panels shared tips on their community-organizing skills, talking about what worked and what didn't.[31]

This youth activism movement is brewing outside the formal school system, yet many students have involved their schools and communities through their excellent convening power.

Youth and Women Empowerment (YOWE)

Youth and Women Empowerment (YOWE), a community-based nonprofit in Ghana's eastern region, has been working since 2000. Its main mission is to support empowerment of vulnerable groups and improve their quality of life through community initiatives, adult learning, and advocacy. Since 2018, YOWE has partnered with the Center for Sustainable Development, Earth Institute, to conduct business entrepreneurial training geared toward eco-friendly upcycled products.[32] The training on green up-cycled products linked to livelihood indicated that sustainability and livelihood can go hand in hand. The inherent tensions between economic opportunities and caring for the planet can be resolved through careful capacity building which brought together existing skills (mainly tailoring) and blending it with eco-consciousness. YOWE's workshops included environmental elements about how they stitched together the water sachets that had been thrown out in the streets. These plastic sachets served as a lining to the cotton tote bags and accompanied by discussions that included the importance of upcycling and the environmental impact of single-use plastic bags. This led to more workshops that helped connect YOWE to local and international markets to provide up-skilling and an economic argument along with eco-consciousness.

Citizen Science

The following example focuses on a citizen science project in rural Madhya Pradesh State in India. Resnik, Elliott, and Miller[33] define citizen science as "a range of collaborative activities between professional scientists and engaged laypeople (citizens) in the conduct of research." Citizen science is a powerful capacity-building tool that has multiple benefits. First, the residents of a location are able to get involved in a local sustainability issue. Second, they can learn scientific procedures and can thus upgrade their skills toward resolving a local problem. Third, the scientists can use the action research methodology to build large databases using crowd-sourced data from the residents to validate their research claims. Finally, citizen science work helps the residents feel connected to a local issue and can shed an investigative light to help in understanding the problem at hand.

In the Alirajpur district of central India, residents primarily depend on government-dug tube wells. As a result, the area has historically recorded high rates of fluoride in the water. High fluoride results from natural causes that include the rocky topography of the region. For that reason, the tub wells dug by the Public Health and Engineering Department (PHED) have a high likelihood of containing more than 1.5 mg per liter of fluoride, which could cause dental fluorosis, weaken bones, and lead to neurological ailments.

In 2018, a group of scientists and social scientists from the Earth Institute conducted a preliminary needs assessment in the district. In the initial needs assessment,[34] the researchers visited community preschool centers, spoke to community health workers, visited district colleges, and found out that there are many rumors and misconceptions about the issue.[35] To work on this issue, the researchers collaborated with a group of sixty social work students from Alirajpur District College, who traveled to villages near their homes to talk about the fluoride issue. Earth Institute researchers then trained the students to conduct fluoride tests, record data on their phones to create a live digital map, color code the tube well according to the content of its fluoride (blue if safe; yellow if high fluoride), and also talk to the village residents and display posters about the ill-effects of fluoride.[36] Using this citizen science approach, the PHED gained support from multiple people who benefited from a defined field-based internship on fluoride awareness, rather than having to rely on its own ten personnel.

The citizen science project also needed intergovernmental partnerships to understand this issue holistically. Therefore, the Education Department also joined hands in a water and health issue through the involvement of college students. The Health Department continued to conduct dental fluorosis detection health camps but with new energy, since the village residents were much more aware. Because the residents had been made aware of the problem by the student interns, they were very receptive to the Health Department when it conducted its health camps. The residents also cooperated with the Water Department to a much greater extent than in previous years, as they had heard the same message from college students from their own village.[37]

There was another element of this study: 60 percent of the student interns were women. This activity was the first time young women had an opportunity to explain a social issue to their village elders and their own family members. They were very hesitant early in the effort but soon realized that the information about dental fluorosis would benefit the rest of the village. They were empowered by their new skill of testing the water from the tube wells, which no one else in their village knew how to do.[38] With this unique skill, they gained new respect in their village, and many people started asking them questions about the water issue.

The study helped everyone involved realize their social capital. The district may have been economically poor, but it is rich in cultural capital. Arts also became a part of the solution;[39] it included street theater to make people aware of the fluoride problem in water. Communicating a scientific issue in an accessible form using local language and arts helped achieve a far wider reach with a long-lasting impact.

On the technical side, PHED adopted the technique used in this study for their work.[40] It helped conduct the students' training and supervised them when Columbia researchers returned. The study highlighted that, first, sustainable development issues are cross-sectoral, and thus various government departments need to come together to solve problems. Second, partnership with ordinary citizens is essential for making any governmental initiative possible. Third, leadership at the local government level is critical to forming collaborations. Finally, local colleges are a great resource, with enthusiastic young students who can take on a societal mission.

BEST PRACTICES IN INFORMAL COMMUNITY-BASED EDUCATION

Often neglected or not discussed, community-based education has the potential to bring about long-lasting impacts in communities, particularly around climate change. This chapter has drawn from many community-driven approaches that won those communities' buy-in from their residents. Through their collective wisdom, these communities observed environmental challenges in their neighborhood and came together to move forward toward finding a solution.

Here we have also provided examples of informal community outreach efforts through educational programs, which can expand through online and offline channels. Technology has aided in outreach and made the community aware of its problem. Often, markets and libraries as common meeting places are not seen as educational places where people of all ages congregate. These meeting spots could be converted into informal educational hubs where environmental movies and posters could be discussed. The educational potential of these public spaces has not been fully explored yet.

Informal education also needs institutional support structures. Governmental departments could come together to support outreach; for example, vaccination campaigns can be combined with environmental messaging. Leveraging existing outreach mechanisms to make them multisectoral will help get the messaging out in a more efficient way. Tapping into environmental commissions, city or town councils, municipalities, and local government structures to include environmental messages could be another way to give climate storytelling the broadest reach.

Community-based education is not lacking in scientific facts; science remains the foundation for environmental messaging. Education programs like Climate Live K12, IRI, and eco-ambassadors invite scientists to discuss changing environmental patterns. Once these facts are received and digested using online and offline media, discussions and message dissemination can create awareness to change behaviors and act locally.

The chapter also discussed how these community-based education systems provide a sense of empowerment through teaching residents skills for solving a sustainability issue that is impacting their lives. Climate activism is empowering. Community members who engage in environmental storytelling and get everyone together to discuss local environmental stories are also community leaders. Their leadership is demonstrated through organizing campaigns, creating messages, connecting with various platforms, and mediating with various levels of the government. They also play a significant role in being involved in scientific research through citizen science, data collection and analysis, and dissemination of relevant climate information that will greatly impact their communities.

The reach of community-based networks is far and wide. Informal education taps into the community's social networks and social capital. This rich network enables the community to be environmentally focused in many ways. It is a great way to connect to elders' Indigenous knowledge, which helps members connect with nature. Here we have illustrated some ways that communities become involved through educational activities.

The systems, processes, and structures within a community are critical avenues for incorporating climate change education. We hope that through these examples, we have provided a glimpse into how communities worldwide are engaging in meaningful discussions and work around climate change. The next chapter will examine climate change education in nonformal learning environments and their essential role in enhancing climate literacy.

6

TEACHING CLIMATE CHANGE IN NONFORMAL SETTINGS

Midway between formal and community (informal) learning lies nonformal learning, which perhaps is a combination of formal and informal learning in that it blends approaches used in both environments. For the purposes of this chapter, we are use the following definition: "Non-formal learning takes place outside formal learning environments but within some kind of organizational framework. It arises from the learner's conscious decision to master a particular activity, skill or area of knowledge and is thus the result of intentional effort. But it need not follow a formal syllabus or be governed by external accreditation and assessment."[1]

We see nonformal learning—sometimes also referred to as "experiential learning"—as a series of activities that are organized and can have learning objectives. It's beneficial to think of nonformal learning as occurring in a space different from formal and informal settings; there are opportunities for teaching and learning in this intermediate space. We've seen from programs such as those described in the previous chapter that individuals often benefit from organized activities in an informal setting, and often important by-products of this learning may not always be obvious.

Nonformal learning environments are important places to consider for discussing climate change and systems thinking. They often represent a place of coming together—a common ground of sorts, where all participants can work collectively toward a particular goal. In formal settings, educators have to adhere to achieving very clear outcomes, and there is less time and space to discuss important systems pieces of climate change, such as politics and emotions.

Climate change has become a politically laden subject, and sometimes it can be easier to not talk about it at all than to discuss it in certain learning environments. This is where experiential settings come into play, unique places where a deliberate focus can be made on strengthening learners' climate literacies but also where learners' understanding of climate can be shaped informally through discussions and activities with others.

Talking about climate change these days inevitably involves bringing up politics and with it, strong emotions. Researchers at Stanford University's Climate Change Education Project have been documenting the linkages among climate change, emotions, and politics since 2009. They have found that feelings of fear, helplessness, and even anger are common among young people when discussing climate change.[2]

In a classroom setting, politics and emotions are difficult to address, but they greatly influence how a student learns and what concepts they grasp. That is why organized activities with clear objectives in an experiential learning environment outside the classroom are critical.

In the following section, we look at case studies that serve as helpful examples of nonformal learning programs that have been developed to meet the need for structured learning opportunities outside the classroom. These programs range in their structure and daily activities, but we can find commonalities in their overall goals, which take aim at integrating science with important social and political histories of climate change so learners develop and strengthen their overall scientific literacy. This ultimately helps students better understand and grasp the complicated dimensions of climate change in both the short and long term.

NEXT GENERATION OF HUDSON RIVER EDUCATORS

Columbia Climate School's "Next Generation of Hudson River Educators" program is a six-week summer internship designed to more effectively engage underrepresented minority students and communities with the Hudson River. It is funded by the Department of Environmental Conservation in New York state and run through Columbia's Lamont-Doherty Earth Observatory.[3] The program follows these themes:

- Week 1: Hudson River Ecology and Transitional Ecology
- Week 2: Science Communication
- Week 3: Environmental Justice
- Week 4: Making Scientific Data Accessible
- Week 5: Interviewing and Learning from Local Communities
- Week 6: Bridging Communities to the Hudson Through Engaging and Relevant Education[4]

The program targets high school students, who spend the beginning of the program learning about the Hudson River estuary through field investigations, building an appreciation for the estuary that they ultimately share with their own communities through materials that they develop. Introducing students to the Hudson River estuary uses learning tools such as videos, games, interactive web activities, and live demonstrations by the water.

As part of learning about the Hudson, students also engage in a process of data collection around the waterfront in Piermont and Haverstraw, gathering information on fish abundance and diversity, water chemistry, habitat assessment, soil chemistry, and lead in the soil. They also assess lead in their own homes and yards. These young scientists not only conducted research on the river but also strive to learn about, personally represent, and reach out to the full diversity of communities along Rockland County's river.

A key part of the program includes interviewing friends, family, and community members to learn about their relationship with the Hudson River estuary to better develop and direct education and communication materials about this critical resource. Conversations with residents in the park and on the waterfront have led the students to develop a wide range of outreach materials. They have partnered with stakeholders to create targeted messages through different media forms for different riverfront communities on fishing restrictions; developed data-rich games that can be shared through outreach events and outdoor science events; created Instagram posts on native versus invasive plant species in their communities; crafted short single-topic videos; and written a blog post discussing topics like environmental justice, the long-term impacts of redlining in communities, sustainable planning, the power of learning from the community voice, and Hudson River science. Through their work, they have developed a network of peers with common interests and a richer understanding of this enormous resource in their very own backyard.

NYC RESEARCH SCIENCE MENTORING CONSORTIUM

This consortium is based at the American Museum of Natural History and is a partnership of academic, research, and cultural institutions committed to providing dynamic, mentored science research experiences for New York City high school students who have an interest in STEM.[5]

The consortium is supported through the Pinkerton Foundation, an independent grant-making organization dedicated to helping young people reach their full potential. The programs that make up the consortium focus on recruiting New York City students in grades ten through twelve and providing them with a unique opportunity to work with scientists on an array of mentored research projects that take place in nonformal/experiential learning settings.

The goal of the consortium is to engage the increasing numbers of underserved high school students—many from groups underrepresented in the sciences—in programs in which they can conduct research working alongside scientist mentors, enabling them to achieve success in college and motivating them to pursue careers in scientific fields. While each consortium member's program is unique, all are guided by the same set of core principles and best practices identified by the program partners. It also cultivates a community of practice among a diverse network of scientists, graduate students, educators, and like-minded individuals. Together, this community invests in and empowers their scholars to succeed in STEM academically and professionally. Next, we describe one example of a program that operates within this consortium.

Woodland Ecology Research Mentorship

Wave Hill's Woodland Ecology Research Mentorship (WERM) is a fourteen-month paid program "offering motivated New York City high school students a unique opportunity to gain in-depth knowledge of ecology and participate in hands-on fieldwork and authentic science research."[6] "WERM interns are given various responsibilities over the course of the program, developing into independent researchers by its end. Each phase is composed of various activities, including hands-on fieldwork, coursework, collaborative projects, and research with scientist mentors." Through this holistic approach, WERM interns learn about urban ecology in an authentic science research context.

In total, the WERM program has three phases over the course of 14 months, with each phase having a series of strategic activities that allow students to do fieldwork, coursework, collaborate with peers, and work with scientists in research settings; here is a detailed overview.[7]

> Phase 1: During their first summer, interns participate in two college-level courses and engage in forest restoration. Throughout the academic year (Phase I), they build their knowledge of New York City's natural areas and learn important research skills through workshops and field trips. During Phase I, interns gain foundational knowledge of the principles of restoration ecology, standard data collection methods, and essential tools for scientific analysis.
>
> Phase 2: In the spring and following summer (Phase II), interns take a third college-level course and work with scientist mentors to conduct research on an ecological topic before presenting their research at graduation. Phase II consists of

"WERMshops," which take place on Saturdays during the school year and also have the opportunity to earn community service hours by participating in ecological monitoring and restoration work after school and during school breaks. Phase II is dedicated to building on what has been learned in the summer—deepening interns' understanding of the various methodologies used in scientific research and knowledge of the ecology of New York City's natural areas.

Phase III, students participate in an immersive research experience. In Phase III, interns spend the majority of their time working on mentor-led small-group research projects. To support students in their research, interns meet weekly to take the course "Introduction to Science Research." In addition, interns participate in an improv class that helps them develop communication and public speaking skills. The phase culminates with interns presenting their research at a celebratory graduation symposium in August.

LABVENTURE

A twelve-year statewide free science program called "LabVenture" has brought middle school students and teachers from Maine to the Gulf of Maine Research Institute (GMRI) to immerse them in hands-on explorations of local marine science ecosystems.[8] The goals of the program are to build climate and data literacy; create spaces where students can build and demonstrate communication, collaboration, problem-solving, and critical thinking skills; and provide opportunities for them to understand and apply key scientific practices. This program has served late elementary and middle schoolers in Maine, and thus more than ten thousand students each year have the opportunity to connect learning from inside and outside school through supporting resources and educator professional development.[9]

The goal of LabVenture is to provide the next generation of science learners with the knowledge and data skills they will need to manage the challenges and opportunities that come with a warming planet. It is a unique science learning experience that places students in the middle of some of the most important questions scientists explore every day, with a focus on their own communities and environments.

The experience combines traditional science tools with high-tech interactive resources to enable students to explore Gulf of Maine ecosystems. Using real data from a variety of sources (including NASA satellites and local fisheries), students investigate how warming ocean temperatures are impacting species. As they work together, key scientific skills are built and lifelong skills (e.g., collaboration, problem solving) are nurtured.

After students leave the interactive learning facility, they remain engaged through curriculum modules and educators who have participated in professional development activities. This extended learning opportunity in a conformal environment gives students a chance to sustain engagement around authentic experiences through strong scientific inquiry.

GREEN HUB FELLOW INDIA

Green Hub, a nonprofit in India, uses digital storytelling as a medium to empower rural and tribal youth.[10] Its vision is to empower young people with knowledge about conservation and climate sustainability using their own villages as a learning medium. They are trained in using digital media as a means for telling local environmental conservation stories.

The participants are enrolled in a ten-month fellowship, after which they are connected to various governmental bodies and businesses as interns to showcase local environmental stories using videos. In many cases, the internship turns into a job opportunity for the fellows.

The Green Hub Fellowship involves learning technical aspects of filming, editing, and storytelling. The program provides an opportunity for the youth to (re)connect to their local villages with a lens of sustainability. Fellows are recruited from remote tribal areas and marginalized communities and are transformed into ecological protectors of their local habitat. The nonprofit began operations in Tezpur Assam in 2014 and has spread to other parts of central India. Green Hub is an excellent example that livelihoods and environmental protection are not at odds, and this is made possible through digital education.

FUTURE COAST

Stephen Siperstein codeveloped a digital climate storytelling project called "The Climate Stories Project" with Columbia University's PoLAR Climate Change Education Partnership project.[11] The highlight of this project, led by Stephanie Pfirman, is its ability to capture storytelling at its best by documenting witnessed accounts of climate change around the world. In this project, students capture both the science and their own personal experiences through publicly shared stories that centers them as storytellers, scientists, and planetary citizens. The stories highlight what students believe their future might be like in an environment that has felt the deep impacts of climate change. Through the use of various media tools, students composed and published "voicemails from the future" on a website.

The voicemails are spoken narratives that imagine local climate shifts from a scientific point of view and couples that with stories about mitigation, remediation, and endurance.[12]

This example perfectly showcases what happens when students are given guidance and prompts that channel their feelings, anxieties, and emotions in response to scientific evidence and turn them into action and agency, as well as enable them to think critically about what our futures might look like. Additionally, the storytelling project gives students the ability to write fictional blueprints for what they want to see in an ecologically and socially sustainable world, and allows them to reflect on their role in such a world.[13] Storytelling opens the door for students and teachers to turn an otherwise gloom-and-doom scientific message into one that reflects resilience and strength.

By bringing climate change content into the realms of social sciences and humanities, we can begin to stitch the pieces together and bridge what our scientific realities mean for humanity. By humanizing hard and often distant science, we can move toward a discussion that includes history and stories and use them to motivate people despite feelings of hopelessness and despair. This example lets us employ the power of words and stories, not just data and graphs, to direct how we teach, learn, and experience climate change. Our lived experiences as human beings need to be incorporated into the narrative if we are to cement ways of thinking and doing that will allow us to address the challenges of our changing climate collectively and effectively.

GORONGOSA CLUBS

Mozambique is home to Gorongosa National Park, a preserved area of more than fifteen hundred miles that focuses on wildlife conservation, ecosystem preservation, and promotion of eco-tourism to benefit local communities. The area boasts several informal youth education opportunities, like clubs for girls and all youth, eco-oriented clubs, and ways to promote park literacy.

Girls clubs provide a safe environment for young women to meet every weekday to study and learn life skills, such as writing and math, while providing school materials, uniforms, and mentorship opportunities. This activity began in 2016 and has expanded to fifty clubs that serve two thousand girls across Mozambique.[14] Participants can go on field trips to learn about career options, and career counseling is available. The program has begun to combat alarming dropout rates among Mozambique girls. These girls may have to walk long distances to attend classes, which may not be feasible, or be told to perform household chores or prioritize other activities, as females are marginalized due to cultural norms in this area.[15] Though 94 percent of Mozambique girls enroll in primary school, only 11 percent will continue on to secondary schools, so these club programs provide quality education opportunities in safe environments that empower them.

Additionally, Mozambique features clubs that target all youth and community education. The youth clubs are for all secondary school youths and center on life skills, leadership, and education about biodiversity conservation and local environmental issues.[16] Other topics of interest can be added to the club program based on what the students would like to learn about.

Eco-clubs are behavior change programs fostering community engagement and behavior adaptation to benefit nature in and around Gorongosa and among the area's people. Finally, the Gorongosa Park Club trains teachers to improve children's literacy with innovative education resources.

These programs aim to supplement whatever education the students receive because enrollment among young women decreases, as does access to education, if the students live in rural communities. Though sixty new schools are being created, there is still inequity in access to schools, and Gorongosa uses the tremendous natural resources available to teach about wildlife, science, and important life skills to youth all over Mozambique.[17]

BEST PRACTICES OF NONFORMAL EDUCATION

The foregoing examples show that the content and structure of the nonformal/experiential education programs differ quite a bit. But some important commonalities are helpful in our discussion of why these particular efforts are important in furthering climate change education for future generations of learners.

First and foremost, guiding the overall programs is a scientific research–based curriculum. The research content varies, but it is grounded in rigorous data collection and analysis, scientific inquiry, and problem solving. The content is also grounded in a local context, which is particularly important for learners. If we are going to engage students in understanding their role in climate change or, more broadly, climate systems, we need to immerse them in learning that is connected to their lived experiences.

The research curriculum builds strong student engagement (both intellectually and emotionally), and learners are embedded in authentic research activities that engage them physically, emotionally, and cognitively. A lack of authenticity subverts engagement; for example, when students are asked to take repetitive measurements just to assimilate techniques. or when the tools applied are incapable of measuring desired quantities to sufficient precision. When the research is authentic and the discoveries are real (and often unpredictable and even uncomfortable), students become excited about their learning. They can also experience failed experiments or disappointing observations, but this disappointment is shared and an important part of the learning process. In a classroom setting, failure is rigidly defined and often determined by test results. In a nonformal setting, important discussions and lessons can be gained from that failure. Therefore, the process of failing but working through challenges is important and often a by-product of experiential learning.

In addition to their scientific content, programs in nonformal settings can further engage students through a social justice lens. Learning in nonformal settings can also be multidisciplinary and is not restricted to siloed subjects. Since the structure is flexible and often connected to real-world problem-solving tasks, transdisciplinary learning is more feasible. Cultural relevance and social justice can also be woven into the curriculum.

The pedagogical practices adopted in these nonformal learning environments focus on a decentralization of the power structure that exists in a classroom between an educator and students to whom knowledge is being imparted. Though differences exist between program leaders and students, the power dynamic is much more fluid. In nonformal settings, there is often more facilitation than direction and dictation. This encourages critical thinking and challenges students to take charge of their learning and making it relevant for themselves. In doing so, their emotions about science and their lived experiences converge, and opportunities are afforded to explore their new conversations and situations. Through science, questions about justice, equity, diversity and inclusion are raised, but learners have a unique opportunity to extend those discussions beyond the scientific context.

Another commonality among nonformal educational programs is a strong component of building a community through team-based learning. Teams provide learners with an opportunity to build a community of practice among very different individuals. Teams are a strategic part of any nonformal program, and leaders can often decide how the teams should be formed. Teams are often a mix of individuals who have different strengths and weaknesses, but all have to learn to work with one another and develop empathy. Team-based learning is collaborative in nature and enables students to feel part of a larger collective effort and understand that their role and work is important as part of a broader continuum. This role mirrors their role as members of society who are grappling with changing climatic conditions. In addition to being scientific partners, team members are social partners, and learning goes beyond understanding and knowledge, with the result that students fulfill their emotional and interpersonal needs within the context of a science learning project.

Finally, a critical component is the physical place in which nonformal learning takes place. It is not the classroom, and regardless of the actual venue, the learner is grounded in an environment that can increase awareness and connectedness to their community. These places are novel and different from what students experience in a school setting, and while such environments can be unfamiliar and daunting at first, students find their way around them and establish a sense of belonging over time. The research and curriculum that guide these experiential experiences help promote this feeling of belonging. When learners feel that they are part of a collective effort, their work is useful, and they are personally contributing to a larger endeavor, they can be themselves and try new ideas. These nonformal experiences provide not only the science but also the emotional, interpersonal, cultural, and physical space that is needed for them to grow as individuals.

Efforts to enhance climate change education typically target schools and focus on such strategies as improving science curriculum and teacher training and strengthening the science pipeline. While those are important factors to consider, we must not overlook or underestimate the potential for science learning to happen in nonformal settings. That is where a lot of students spend a significant chunk of their time.

At the same time, we must remember that nonformal and experiential programs should not just replicate what is done in the classroom. While there is a need for content structure, grassroots and sometimes ad hoc activities can prove to be extremely beneficial. The focus in nonformal settings should not be to align activities done in these unique settings to standardized testing requirements in schools. Instead, the idea is to facilitate scientific skills and understanding in ways that are more appropriate for the various settings and activities, as well as to promote interest in science topics and a person's self-identification as someone who is knowledgeable about science.

During nonformal and experiential learning programs, learners have a greater say in what they will learn and how they learn it. Make no mistake—there are often strong feelings behind those intentions, and that's important to remember. When students feel like they can belong and are welcomed in a space, personal growth can happen in a different way versus in a formal classroom. The connections learners make in nonformal settings provide us with an important lesson that is important for climate change education: when students are assisted in creating a culture of well-being through camaraderie, in which their different backgrounds and histories are valued, only then can all of us—students and mentors together—discover that science is for everyone.

As we move forward with efforts to bring systems thinking and climate change education into more learning environments, it's important to remember the importance of the variety of types of learning opportunities that can be offered to learners of all ages, as well as educators, through a variety of environments. In the last three chapters, we covered formal, informal, and nonformal education, and all undeniably important and connected as we move forward. Collectively, climate change educators and advocates need to plan for and expand opportunities in all learning environments, particularly for students from communities underrepresented in the science fields (more on this in a later chapter).

Regardless of the environment—formal, community-based, or nonformal—overarching goals for providing climate change education should be established. Those goals include designing and delivering professional development to support educators, building stronger links between and across all types of learning institutions, improving systems for measuring and assessing all types of learning experiences, and providing greater recognition and support for all those working to achieve stronger science literacy for future generations. By making stronger connections between different learning environments and institutions, we can complement student learning through a variety of interactions and discussions and provide learners with the capacity to build on their knowledge through a variety of means. Each learning environment should operate in the manner of a system, just like the content, to support intentional learning that goes beyond a traditional model of education whereby students learn in a one-directional way by gaining knowledge from an educator.

Learning across these three contexts can be purposeful and enables learners (both teachers and students) to gain knowledge by doing and exploring various educational pathways that pertain to their interests. By better connecting the goals and outcomes across different learning environments by grounding them in systems thinking, educators can support learners by adding to their knowledge base and their toolbox for translating ideas into practice and actions.

IV

THE FUTURE OF CLIMATE EDUCATION

7

DIVERSITY, EQUITY, INCLUSION, AND ACCESS AS A TOOL FOR ADDRESSING SOCIAL AND ENVIRONMENTAL JUSTICE

Now that we have looked at the various types of learning environments and how and why each provides new and unique ways of teaching and learning about climate change, we shift to another big-picture goal: ensuring that regardless of where such teaching and learning occurs, we need to be aware of the socioeconomic and environmental justice challenges that pervade climate change. At the heart of this issue is the reality that communities that are most affected by the uneven impacts of climate change are often the least represented in climate change education efforts. We need to be explicit and strategic if we are to begin to bridge social and environmental realities with science, as they are inextricably linked.

In the past twenty years, the topic of environmental justice topic has sprung up to catch the attention of all. The term that has been traditionally used is "environmental justice," not "climate justice," and refers to equal treatment of everyone—irrespective of their race, culture, and income—in the development of environmental laws, regulations, and policies.[1] Since the 1980s from very modest grassroots beginnings to the present day, environmental justice has come a long way. Unfortunately, its full potential is yet to be reached. As the number of disasters has increased and intensified over the years, a pattern has emerged. Economically and socially marginalized communities have borne most of the brunt of climate change.

Climate disasters impact communities in numerous ways. Many people are forced to migrate, leaving their jobs and seeking new ones; children are displaced; increased health risks are prevalent as extreme heat rises; and often there are no safety nets to provide adequate compensation and support for the economic and emotional losses. As a result, marginalized communities do not get the support they need for educational purposes either. A different set of challenges prevent the same marginalized communities from accessing equitable learning opportunities and quality education for current and future generations. The roots of these challenges are similar in the sense that structural barriers fueled by racism and the practices of "othering" continue to keep marginalized communities in the margins.

As a result, the same marginalized communities that are at risk of bearing the brunt of the impacts of climate change are also the ones that are not well represented in the STEM or environmental science fields. The disconnect to climate change education is often even more significant. The disengagement in STEM in the United States among women, Latinos, and African Americans is higher than in other countries, and the pursuit of secondary degrees in those fields is also low among those groups.[2]

Throughout this primer, our goal has been to focus on the importance of bringing systems thinking and climate change education into learning environments, but that goal is also not free of the racism embedded deep within our society. Educational inequities are pervasive, and the same groups that are most unevenly impacted by climate change are the ones who face the most opportunity gaps in participation in the scientific process. In this concluding section of the book, it's important to be mindful of the racial injustices that pervade educational systems, particularly in the United States.

This primer addresses the importance of educating future generations about climate change to facilitate a greater understanding of our world as a system and encourage young learners to engage in that process. But we cannot separate this goal from important diversity and equity challenges that persist in the climate science and education field.

The reality is that many historically underrepresented groups, such as women, ethnic minorities, students from low-income households, and students with disabilities, are the least likely groups to participate in this educational process. To name just a few of the challenges, the schools these groups attend are often underfunded; there is a smaller network with social capital around them to support their educational efforts; and they are afforded fewer opportunities to participate in enrichment opportunities to support their interest in science learning. These barriers influence how they ultimately engage in science regardless of the learning environment. And even for individuals from underrepresented groups who do engage, the challenges do not stop. The lack of a sense of belonging, as well as racism, sexism, cultural differences, and imposter syndrome are some of the many barriers preventing people from remaining engaged with climate science and science education.

As a concrete example, numerous programs we have run at our institution in an effort to broaden participation in the STEM fields have faced challenges. The integration of students of color into a largely white, upper-middle-class, older community has not been without struggle. Through these experiences, we have seen clearly that there is a wide range of sensitivities to the racial history of America. And while generally we may be supportive of broadening participation in the STEM fields, we often do not practice what we preach. Many colleagues do not see or understand how institutional structures and barriers operate in a way that may affect persons of color, nor do they realize that what they believe are race-neutral comments are not.

While past and current world events show us clear linkages between the environment and social and economic development, their integration into educational content and curricula have been slow to appear. And even for the content that has been developed, we need to be mindful of who has created the content, whose voices the content draws on, and how we effect change so the process of creation becomes a joint effort rather than being dominated by one narrative. While we support the incorporation of climate change content into a variety of learning environments, we need to be aware of how that information was integrated and whether or not it makes important connections between science and our society at large. Climate change science—in fact, all science—is often seen as an objective and factual field. However, that's not always the case. Those who typically dominate the conversations and drive the research forward are often those in powerful positions that are not accessible to all voices. While we'd like to think that our science is objective, it is often laden with injustice.

The intersection of many factors creates sustained and conscious injustices within society. In addition, disasters such as hurricanes also impact marginalized communities disproportionately and compounds existing challenges that often keep these families in a vicious cycle of managing one traumatic event after another. Therefore, many climate activists have equated climate justice to social justice, and we should too, both in our ongoing work in the scientific community and within our educational systems. Open participation and communication in decision-making is the first step toward being ensuring that policy planning and implementation are inclusive and oriented toward climate justice. Critically analyzing what is excluded in school textbooks and why will help make the dialogue more inclusive. Cultures, histories, and contexts need to be considered in the treatment of climate education in schools and other settings.

In the next section, we discuss the trajectories of climate justice in the United States and globally and show the need to look at structural inequalities in climate education.

TRAJECTORIES TOWARD CLIMATE JUSTICE

We are writing this primer in a time when George Floyd was murdered; the long-lasting effects of the COVID-19 pandemic have had a disproportionate impact on African Americans, Latinx, and indigenous communities; and the pandemic has fueled hate crimes against Asian Americans. At first glance, these events may not have the most obvious links to climate change. But if we look at the issue a bit deeper, we can see that racism remains deeply embedded in our society, and it is as present in our environment as ever. Both historical and present-day injustices have left people of color exposed to far greater challenges than whites, whether in regard to health care, education, or environmental hazards.

A recent report by the Sustainable Development Solutions Network about how well U.S. states have met sustainable development goals, pointed out that none of the states is on track to meet the goals by 2030.[3] The main reason is that many communities of color—especially Asian, Black, Indigenous, Hawaiian and Pacific Islander, Hispanic, multiracial, and "other" are not being engaged in important conversations. The groups and individuals who are being left out are the same who have the least amount of access to social services, often live in substandard homes in redlined districts, and ultimately are the ones who will face the greatest environmental threats.

It is important to recognize that the fight against climate change is deeply intertwined with racial injustices in this country. Whatever goals we hope to achieve in the future when it comes to climate change, we need to be aware of how our past actions have worsened the current condition for many. When we think about the current environmental movement, we must understand that it was created by people who prioritized conservation but did not include diverse voices in the process and conversations.

The United States is not the only country where minorities have been ignored in the climate justice dialogue. Next, we present an outline of a brief history of important landmark moments that have shifted the global discourse from being exclusively about environmental conservation to addressing climate justice. Many of these global events sparked the sustainable development agenda and the systems connections among social, economic, and environmental factors. Some of the early international conferences and reports are included in table 7.1.

Many natural and human-made disasters were taking place concurrently to shape the world's thinking about the environment. Some of these global events are listed in table 7.2. The table includes some policy measures that were adopted to showcase the multidimensional impact of climate disasters that needed to be tackled by adopting multipronged strategies. The events show the extent of damage to human life as a result of

TABLE 7.1 EARLY CONFERENCES THAT SHAPED SUSTAINABLE DEVELOPMENT

Year	Event
1960s–70s	Early discussions of sustainability about human–environment interaction concerns: "silent springs," limits to growth
1980	First important use of "sustainable development" in the *World Conservation Strategy* by IUCN
1986	Conference on Conservation and Development in Ottawa; defined "sustainable development"
1987	"Our Common Future" (Brundtland Report) popularized the term "sustainable development"

TABLE 7.2 MAJOR GLOBAL EVENTS THAT SHAPED CLIMATE JUSTICE DISCOURSE

Year	Event
1973	Implementation by the United States of legal protections for its heritage in fish, wildlife, and plants
1984	Drought in Ethiopia
1984	Bhopal gas tragedy
1985	Detection of the Antarctic ozone hole
1986	Chernobyl nuclear station accident
1989	Dump of eleven million gallons of oil into Alaska's Prince William Sound by the *Exxon Valdez* tanker
1991	Collapse of Canada's east coast fishery due to biomass; increasing severe weather conditions and environmental disasters
1994	Release of China's Agenda 21
1995	Execution of Ken Saro-Wiwa, Nigerian writer, television producer, environmental activist

these disasters, the need for human rights to be protected using legal means, and the rise of climate justice.

Many of these events produced judicial proceedings, and environmental justice was front and center in all of them. The connections between environmental issues became transparent as the global conclaves moved toward sustainability as the driving force. The human impact of environmental disasters (human-made or natural) became very evident and needed to be addressed globally. As part of any climate justice movement, we cannot separate race and the environment. To address these problems, we need to address the systemic racism at the root of it all.

The conferences noted in table 7.3 added equity and empathy to the debate, along with governance and accountability. As a result of the recognition of environmental and climate justice, the idea of sustainable development came to be. Sustainable development represents a way of thinking that is driven by a

TABLE 7.3 KEY INTERNATIONAL CONFERENCES THAT SHAPED SUSTAINABLE DEVELOPMENT

Year	Major summits
1992	UN Conference on Environment and Development (Earth Summit) in Rio de Janeiro; Agenda 21
2002	World Summit on Sustainable Development in Johannesburg; commitment to "building a humane, equitable and caring global society, cognizant of the need for human dignity for all."
2012	Rio+20 UN Conference on Sustainable Development in Rio de Janeiro; merging of "the social, economic and environmental dimensions of sustainability." It also emphasized good governance in meeting sustainable development.

systems thinking approach to problem solving. It is a comprehensive approach that is defined by four pillars:

1. economic development (including the end of extreme poverty),
2. social inclusion,
3. environmental sustainability, and
4. good governance (including peace and security).

This systems thinking approach was a departure from that of the previous decade, which was focused entirely on environmental conservation. As a result of the examples mentioned earlier, we can clearly see the connections among droughts, climate refugees, and oil spills, all of which resulted from a push for economic growth. The four pillars helped reveal these connections. The world also observed that climate catastrophes had the greatest impact on the most economically marginalized people of color, people in the global south, and specific subsections of the population. As the sustainable development agenda was developed, it included the multidimensional aspect of sustainability and thus, by definition, also included the intersectionality that those seeking to add the justice component had hoped for. For instance, the good governance pillar included public participation and government accountability as the core of environmental justice. The social inclusion part of the definition of environmental justice states that fair treatment should be given to everyone irrespective of their social and economic status. Therefore, the sustainable development agenda became an overarching umbrella that could comprehensively include the justice arguments.

The framework of sustainable development encompassing the four pillars was more comprehensive than the environmental conservation focus of the previous decades. With the emergence of the sustainable development agenda, justice issues found a reference and a place more so than before. However, often frameworks do not lead to prioritization in real life. In reality, climate justice seems to be marginalized in curricula and classroom practices.

CURRICULAR ANALYSIS WITH ENVIRONMENTAL JUSTICE AS THE LENS

There is a need to discuss environmental justice in various learning environments, however, environmental justice issues largely go unnoticed by curriculum developers and education systems.[4] Kushmerick, Young, and Stein conducted an analysis of U.S. environmental education curricula for grades six to twelve through an environmental justice lens.[5] They included 224 lessons as their data for analysis. The analysis concluded that the curriculum often discussed issues that are tangential to environmental justice, such as environmental health impacts on humans, but rarely did they address topics traditionally considered as justice issues. The authors suggested that there have been many missed opportunities to address justice issues in curricula and textbooks.

Kushmerick et al.'s curricular analysis reaffirmed Lewis and James's observation that although the theoretical foundations of environmental education are inclusive, the practical applications of this framework in school curricula are hard to find. Lewis and James argue that discussing the great impact of environmental damage is irrelevant for students unless its source and the social, economic, and political dimensions are also discussed.[6] Their analysis revealed many environmental justice terms used in the curriculum, but the content taught explicitly about environmental justice was weak.

FINAL THOUGHTS

Unless curricula, schools, communities, and policies provide considerable space to explore the systems connections that fuel the social inequalities and advance the dialogue and reflections on environmental justice, structural factors will continue to widen them. These structural inequalities need a place in the curriculum for critical analysis leading to action that brings system-wide change. The topics and content need to be included in the curricula of all types of education through lifelong learning approaches. Gender-based violence, economic inequalities and marginalization, polarization, racial violence, unequal treatment based on color, religion, and race, as well as environmental injustices are vital topics that any social justice curriculum should include. Ensuring environmental and climate justice starts by recognizing that environmental damage is socially distributed along lines similar to other social and economic disadvantages. The injustice of such distributions has also led to diversity, access, and inclusion issues in STEM fields.

A one-size-fits-all approach is not sufficient for addressing the challenges laid out earlier (and many more that have not been addressed here), which often involve engaging in a long and difficult process of changing cultures within institutions. Because so many factors contribute to systemic racism, gender disparity, and a lack of diversity in the science fields, we need multiple avenues, or a systems approach, to correcting the injustices.

Together, we must identify and mitigate barriers to entry, persistence, success, and advancement for underrepresented and marginalized groups in our science spaces and places, and we must also build equitable practices (by leveling the playing field) into all processes (e.g., hiring). We need to develop resources to help expand perspectives about how to achieve a truly equitable, inclusive, and diverse approach to strengthening climate change education. We also need to amplify authentic and diverse voices and highlight strategies that have worked.

We need to move forward with a systems approach in which we simultaneously address institutional and structural barriers built into current educational systems and ensure an all-hands-on-deck approach as we reimagine the structure and function of the climate science field in the future.

8

ROLE OF THE COLUMBIA CLIMATE SCHOOL IN CLIMATE EDUCATION

As we witness the disproportionate impacts of climate change, we know that collective action is required in order to address climate change from multiple directions. We need to change the decision culture at various levels (e.g., boardrooms, classrooms), and one of the most effective ways to do that is to educate and train future generations to be prepared to tackle the climate crisis in a variety of ways.

Through integrated learning and teaching opportunities, we need students to begin to understand at a young age that climate change is a systems problem. As they join the workforce (not necessarily as scientists), we want them to carry with them multidisciplinary, transdisciplinary, and multiple-perspective thinking to create a workforce that understands the global crisis but can implement local and global strategies.

To accomplish that, we need to ensure that a systems approach to climate education is occurring at all levels from early in students' lives—not just in science classrooms but across the curriculum in formal and informal learning environments. At Columbia University and other academic institutions around the world, one way to make important strides toward achieving this goal through engagement with K–12 stakeholders. With this engagement, we have an opportunity to train and prepare future generations of learners to develop an agile growth mindset through science that can embrace lifelong learning outside classroom walls, to become comfortable with complexity, and to solve cross-cultural problems through teamwork. These climate change educational opportunities provide learners with the capacity for creative invention and imagination, the ability to envision what is not there, and the drive to go forward with confidence.

For too long, studying climate change was seen as simply a matter for physical scientists. In many of our university educational systems, climate has been mostly an afterthought or add-on to most other degree programs. But as we see and feel the effects of climate change more and more, we need to prepare students to grapple with how climate will transform their fields regardless of what they are studying.

Across all levels of education, students don't usually learn about climate change unless they are in a specific degree program that focuses on it. Even so, many students don't learn about climate change and earth systems in their degree programs, and hurdles remain to delivering the kind of universal climate education requirements that are needed. Most recently, academic institutions have offered courses dealing with the physical environment and/or sustainability, but we need more. The connections are so clear that climate change will permeate every aspect of life, and as Earth's systems are changing rapidly, we need our educational programs to keep up.

Academic institutions play an important in ensuring that we reorient places of higher education toward systems thinking for sustainability across all programs. Rather than focusing on climate change just in the science department, we need to be emphasizing climate literacy and systems thinking in all departments and creating guidelines for students to learn and understand how their fields are going to be impacted by wider environmental challenges.

Because we have become so good at specializing in our respective fields, we tend to understand small pieces of the picture extremely well but fail to prepare students to connect their specialties to the larger-picture issues and boundaries beyond their own disciplines. The challenge of incorporating systems thinking into all fields has to be met if students are to be prepared for careers and life in an era of climate change.

The challenge of bringing climate change education with a systems thinking approach is too big to solve alone. Schools and school districts are the places to start planting the seeds of this change, but more is needed. This is where universities, as existing institutional resources, are ideally suited to take on the role of working with schools and school systems to advance climate change education, translate the science of climate change, and make scientific knowledge more accessible to allow for deeper learning.

Of course, educating broad audiences requires other stakeholders as well, such as government organizations, the media, and the private sector. Universities, however, are in a unique position, as they play a large role in educating professionals who work in the aforementioned industries and, like schools and classrooms, have a significant multiplier effect over time. Unlike K–12 formal educational environments, universities are also uniquely positioned to generate content that can feed into curricula for climate change education, as there is strong research and expertise across disciplines, which, combined, can drive not only learning about the science of climate change but also learning that emphasizes the understanding of climate change in the broader context of sustainable development.

As noted, not just science departments can become involved in this important work. Drawing on collaboration across disciplines, universities can develop resources that can be used as important curricula in all learning environments. Universities have the greatest capacity for this kind of interdisciplinary collaboration in developing a rich curriculum and transformative pedagogies that support deeper learning about climate change. Like schools and classrooms, institutions of higher education are also ubiquitous, and therefore the reach and possibilities for advancing climate change education efforts in different contexts are strong.

The need for universities to adapt—and fast—was one of the factors that shaped the motivation to create the Columbia Climate School. Universities play a major role in the education of responsible and informed citizens, and as a result of rapidly changing workforce needs, the school can address those future needs in an important way.

Columbia Climate School's motto is "Climate, Earth and Society" because, in addition to the critical focus on climate, the school will also include the broader issues of sustainability, environment, and populations and their livelihoods. Climate is an extremely broad topic; it intersects many aspects of sustainability, including biodiversity loss, air pollution, disaster relief, food security, and migration These related sustainability and geoscience topics should be included in curricula, not only because the subjects are closely related but because the discipline are as well.

The Climate School now encompasses the Earth Institute and its many research centers and programs. Although relevant work is occurring across the university, the largest part of Columbia's footprint in climate research sits within the Earth Institute's research centers.

The Climate School will be the preeminent institution for climate information and training programs, bringing Columbia's resources, knowledge, and capacity to the rest of the world while simultaneously eliminating the perpetuation of existing inequities. The mission of the Climate School's educational programming (in degree and nondegree offerings) and public outreach is to build climate literacy, leadership, and capacity in individual institutions and enterprises across sectors and communities, infusing climate consideration through a lens of equity into decision-making at every level.

EDUCATION AT THE COLUMBIA CLIMATE SCHOOL

The Columbia Climate School aims to serve as a hub for any learner interested in climate, whether they are in one of the Climate School academic or outreach program or another school's program, or they just want to complete a single elective, attend an event, or participate in a club. Bringing together these students across Columbia and beyond enhances the educational experience of for all and generates an intellectual energy that reflects the vast range of interests of our community.

Under our degree programs, the Climate School is currently expanding its one-year climate and society master's program. The school will also seek to develop a master of science degree program in climate to sit alongside and complement the master's in climate and society. The master of science program has been designed around interdisciplinary core competencies for all students and enables students to specialize through a modular approach.

The creation of additional new master's degrees will be considered and evaluated over time. New degrees will be developed to fill educational gaps, reflect the changing nature of the problem, and meet a need in the market, with a variety of flexible iterations. In many cases, the Columbia Climate School will partner with other schools to offer joint or dual degrees, certificates, or concentrations and specializations, as appropriate.

In addition to the school's degree offerings, we believe there is a fundamental need for nondegree education efforts. The Columbia Climate School's professional learning programming is guided by the principle that channeling knowledge derived from basic and applied research, contributing to the development of an adaptive workforce in a changing landscape, motivating policy and the allocation of public and private resources, aligning institutional decision-making and social capital across sectors, and substantiating individual action are all part of the school's contributions to broadening our impact beyond our own university's walls. The school is also investing in professional learning and precollege programs in climate literacy for participants ranging from primary and secondary school teachers to corporate and nonprofit leaders. These programs and partnerships serve a fundamental purpose in enabling the Climate School to engage beyond the university with governments, the private sector, advocates, nongovernmental organizations, and lifelong learners, with the central tenet that such interactions enlarge traditional research and educational programs through collaborations. The particular mission of the Climate School's executive education, partnership initiatives, and public outreach efforts is to build climate literacy, leadership, and capacity in individual institutions and enterprises across sectors and communities, infusing climate considerations through a lens of equity into decision-making at every level. Beyond meeting immediate workforce needs, the school's educational efforts outside of degree offerings democratize access to Columbia.

Engagement through numerous educational offerings allows us to step away from our ivory tower. It also offers learners of diverse backgrounds a variety of touch points and opportunities to engage with our important work.

Colleges should no longer be merely places you go to earn a degree. The Columbia Climate School can foster, facilitate, and grow greater engagement with broader audiences with an eye toward lifelong learning. The school can play a key role in extending our abilities beyond our academic community around the pressing challenge of climate change.

CONCLUSION

We'd like to offer some concluding thoughts to consider as we move forward with future efforts around educating learners about climate change, systems thinking, and justice. Our motivations in producing this primer have revolved around the challenge that many learners are not being adequately educated about climate change despite the scientific evidence and understanding that has come about. While education in this area is acknowledged as important, implementation and how it is addressed in learning environments of all kinds varies greatly.

Our educational system is out of sync with the scientific community, which has been sounding the alarms for a long time. As a result, in general, understanding of the systems impacts of climate change is lacking and with it, an inability to recognize the urgency of this crisis. What students are learning in science curricula is not synced with the impacts of climate change that is being felt in all sectors.

Another significant social factor affecting climate change communication is political controversy; climate change has become an increasingly polarizing issue in the last decade. Political controversy affects climate change at the state level, where policy makers dispute the content of educational standards, as well as the local level, where teachers, administrators, parents, and students must negotiate whether and how to include this topic in schools. Even in communities where climate change is not particularly politically divisive, other social and emotional factors—such as students' feelings of hopelessness or a teacher's fear of instigating conflict—may further complicate climate change in the classroom.

We, along with many others in this field, have hoped to convey that education has continued and will continue to play a huge role in preparing future generations to address and mitigate the effects of a changing climate. If we continue to educate students about climate change in a silo and without a systems approach to learning and teaching, we will leave them unequipped to make future decisions when they enter the workforce and become decision-makers in different industries. When we fail to teach learners how to make connections between what they are taught in schools and their lived experiences, we fail them, and we are not giving them the whole truth.

We have made a deliberate choice to focus on education as a potential tool and method that we will use to break down the hegemonic structures that have so far been barriers to climate change education. Climate change has indeed been heavily politicized, science itself is under attack, and science innovation is often defined by those with the power and presence to do so. But what we have realized through our own work is that education and preparing learners with important twenty-first-century skills (such as critical thinking, problem solving, and collaboration) is an issue that everyone, regardless of race or political beliefs, can rally around.

Our classrooms and schools should not be battlegrounds for climate change; they should be safe places for learning and teaching that provide students with the opportunity to discuss complex problems, ask questions, and work through difficult solutions. And even when climate change is discussed in classrooms, it is also important to remember that the details of how these systems concepts are taught really does matter.

So we are championing a different approach to bringing sound science into learning environments. We want to focus the narrative on equipping the youth of today with the skills that will enable them to join and add value to future workforces and address climate change challenges through a systems approach. Therefore, we have highlighted the importance of systems thinking, because the most effective climate change education program or curriculum will be the one that makes connections between global climate change processes and their local impacts and effects. The most effective educational content will also be transdisciplinary and incorporate not only new scientific discoveries but also historic ecological knowledge and indigenous knowledge to understand our past and present in order to prepare for our future. Many individuals and groups are working hard to reverse the alarming climatic trend that we've seen over the past decade, and as that important work continues, we want to be sure to emphasize the importance of learning and teaching opportunities outside schools and classrooms.

We've provided a few examples of out-of-classroom learning environments that have been successful in teaching about climate change and earth systems science, but not mentioned here are many other groups (e.g., at museums and parks, in faith-based organizations) that are also engaged in this important work. If we are to achieve a future workforce that is truly more climate systems–literate, we need to promote these educational efforts across all subject areas from early in learners' lives and in multiple learning environments, so that if one space becomes politically charged, teaching and learning can continue in another. All educators, parents, and even the general public have a role to play in shaping the discourse around climate change and for preparing students for the climate crisis that we have begun to experience. It is vital that we continue to work toward climate change education *for* all and *in* all learning communities, because what we do (and don't do) now will affect how climate change impacts will be felt and dealt with in the future.

While the challenge is complex and there is great urgency, we should not rely on scare tactics for educational purposes. Systems thinking in climate change education should not involve the teaching and learning of just the problems. We need to ensure that solutions, mitigation, adaptation, and the role of youth are important components of any educational program. Instead of scaring learners into becoming interested and acting, we should be encouraging, facilitating, and energizing future generations to take on climate change in ways they are familiar with. To adapt to a warming planet, we need all hands on deck, and by creating an inclusive and diverse community early on, we can be sure to leverage our collective strengths from the get-go.

Science, particularly knowledge generated from postsecondary institutions, remains an important and trusted source of information. But there are inevitably gaps. Often, science content is not always accessible to broader audiences, including school districts, teachers, and students. Therefore, we need to do a better job of making those linkages with school systems and ensuring that the knowledge transfer takes place in an effective way so everyone can join the conversation and be engaged in the content.

Looking forward, we also need to ensure that the science and solutions behind the climate crisis are not a body of knowledge that students merely memorize, reproduce for a test, and never use again. We want climate change education to do the exact opposite. We want students to make the important systems connections about how climate change processes happen globally but have impacts that are felt locally, that the changes will absolutely have great relevance for their personal lives, and that the environmental, social and political impacts will have immediate real-world implications for them and their loved ones.

If we are to achieve this goal, we need to be doing more to set up learning environments where students are taught to use the knowledge they gain about climate change to contribute to their communities, become responsible citizens, and take action in meaningful ways. These outcomes are possible only if a system thinking approach to climate education is adopted.

We need to ensure that the teaching and learning that occurs in a variety of environments are helping students make the explicit connections between global climate change and locally felt effects and impacts. For example, during the COVID-19 pandemic, online learning changed drastically and provided a new avenue of educational content delivery that is now a part of how we bring education to the masses.

We should be creating engaging experiences for learners through transdisciplinary content that bridges the physical and social sciences; at the same time, these experiences need to enable students to discover different ways to engage with the content in different contexts. Whether students learn through labs, field trips, or scenarios and simulations in or out of the classroom, the key is to create opportunities for learners that meet them where they are and leverage their strengths to further their personal as well as collective learning.

We also need to take into account the different types of knowledge that are out there. Scientific knowledge produced by research universities is not the only source of information.

Incorporating indigenous knowledge that has been passed on from generation to generation long before schools were ever set up is one example of connecting history and science. For example, Indigenous communities have long practiced sustainability, and it is deeply ingrained in their ways of life. Living in harmony with the Earth to create something of use without changing any of the fundamentals is not an idea that has surfaced recently. In fact, these practices and ways of life should be highlighted as profound insights into how we approach peaceful coexistence with our environment. Important knowledge generation can be found in nonformal settings, and it is our duty as educators to connect learners to all educational avenues beyond classroom walls.

Finally, we should be learning from and incorporating the diverse approaches of Indigenous communities for our science, our education, and beyond. We've seen what happens when there is an imbalance and we lose our sustainable relationship with nature. We need look no further than the Industrial Revolution to see that less than two hundred years after that period of tremendous economic and technological gains, we are facing grave environmental issues resulting from it.

This idea of achieving and maintaining balance is at the heart of systems thinking. Everything is deeply connected, and there is a delicate balance in maintaining a sustainable system. To address that interconnectedness, we need to be utilizing a variety of methods of teaching and learning. A transdisciplinary approach to climate change education through systems thinking will allow important interconnections between humans and our environments to be recognized through place-based, observational, and participatory methods that are inclusive and meaningful for all. Ultimately, that is what's needed to reach a balanced and just society in which sustainability for all can be achieved.

NOTES

FOREWORD

1. Katie Worth, *Miseducation: How Climate Change Is Taught in America* (New York: Columbia Global Reports, 2021).

INTRODUCTION

1. See details at "Education for Sustainable Development," UNESCO, https://en.unesco.org/themes/education-sustainable-development/what-is-esd.

1. DEFINING SYSTEMS THINKING AND CLIMATE CHANGE

1. United Nations Educational, Scientific and Cultural, Organization (UNESCO), *Education for Sustainable Development Sourcebook* (Education for Sustainable Development in Action, Learning & Training Tools No. 4, UNESCO, Paris, 2012), http://unesdoc.unesco.org/images/0021/002163/216383e.pdf.
2. Matthew T. Ballew et al., "Systems Thinking as a Pathway to Global Warming Beliefs and Attitudes Through an Ecological Worldview," *Environmental Sciences* 116, no. 17 (2019): 8214–219.
3. B. S. Bloom, M. D. Englehart, E. J. Furst, W. H. Hill, and D. R. Krathwol, *Taxonomy of Educational Objectives: The Classification of Educational Goals, Handbook I: Cognitive Domain* (New York: David McKay,

1956); and B. S. Bloom, *Human Characteristics and School Learning* (New York: McGraw-Hill, 1976).

4. P. G. Schrader and Kimberly A. Lawless, "The Knowledge, Attitudes, and Behaviors Approach: How to Evaluate Performance and Learning in Complex Environments," *Performance Improvement* 43, no. 9 (October 2004): 8–15.
5. I. Ajzen and M. Fishbein, "Attitude-Behavior Relations: A Theoretical Analysis and Review of Empirical Research," *Psychological Bulletin* 84 (1977): 888–918.
6. M. S. Kim and J. E. Hunter, "Relationships Among Attitudes, Behavioral Intentions, and Behavior: A Meta-Analysis of Past Research, Part 2," *Communication Research* 20 (1993): 331–64.
7. K. A. Lawless, J. M. Kulikowich, and E. V. Smith, "Examining the Relationships Among Knowledge and Interest and Perceived Knowledge and Interest" (paper presented at the annual meeting of the American Educational Research Association, New Orleans, LA, April 2002).
8. C. Byrd-Bredbenner, L. H. O'Connell, and B. Shannon, "Junior High Home Economics Curriculum: Its Effect on Students' Knowledge, Attitude, and Behavior," *Home Economics Research Journal* 11, no. 2 (1982): 124–33.
9. P. A. Alexander and T. L. Jetton, "Learning from Text: A Multidimensional and Developmental Perspective," in *Handbook of Reading Research*, vol. 3, ed. M. L. Kamil, P. B. Mosenthal, P. D. Pearson, and R. Barr (Mahwah, NJ: Erlbaum, 2000), 285–310.
10. P. A. Alexander, "The Development of Expertise: The Journey from Acclimation to Proficiency," *Educational Researcher* 32, no. 8 (2003): 10–14.
11. G. W. Allport, "Attitudes," in *Readings in Attitude Theory and Measurement*, ed. M. Fishbein (New York: Wiley, 1967), 1–13.
12. L. L. Thurstone, "The Measurement of Social Attitudes," *Journal of Abnormal Social Psychology* 26 (1931): 249–69.
13. Schrader and Lawless, "The Knowledge, Attitudes, and Behaviors Approach."
14. Schrader and Lawless, "The Knowledge, Attitudes, and Behaviors Approach."
15. Barry G. Rabe, Christopher P. Borick, and Erick Lachapelle, *Climate Compared: Public Opinion on Climate Change in the United States and Canada* (Center for Local, State, and Urban Policy [CLOSUP] Report,

Brookings Institution, Washington, D.C., February 1, 2011), https://ssrn.com/abstract=2313303.
16. Rabe et al., *Climate Compared.*
17. National Aeronautics and Space Administration (NASA), *Earth System Science: A Closer View* (report, NASA Advisory Council, Earth System Science Committee, NASA, Washington, D.C., 1988).
18. NASA, *Earth System Science.*
19. S. B. Wise, "Climate Change in the Classroom: Patterns, Motivations, and Barriers to Instruction Among Colorado Science Teachers," *Journal of Geoscience Education* 58, no. 5 (2010): 297–309.
20. UNESCO, *Education for Sustainable Development Sourcebook.*
21. A. Mogren, N. Gericke, and H.-A. Scherp, "Whole School Approaches to Education for Sustainable Development: A Model That Links to School Improvement," *Environmental Education Research*, 25, no. 4 (2019): 508–31.
22. Sustainability and Education Policy Network (SEPN), *Responding to Climate Change: A Primer for K–12 Education* (Sustainability Education Research Institute, Saskatchewan, Canada, 2021), https://sepn.ca/wp-content/uploads/2021/01/SEPN-CCEd-Primer-January-11-2021.pdf.
23. R. A. Duschl, H. A. Schweingruber, and A. W. Shouse, eds., *Taking Science to School: Learning and Teaching Science in Grades K–8* (Washington, D.C.: National Academies Press, 2007).
24. G. King, M. Tomz, and J. Wittenberg, "Making the Most of Statistical Analyses: Improving Interpretation and Presentation, *American Journal of Political Science* 44, no. 2 (2000): 347–61.

2. SYSTEMS THINKING SKILLS AND OUTCOMES

1. R. Hipkins, "More Complex Than Skills: Rethinking the Relationship Between Key Competencies and Curriculum Content" (paper presented at the International Conference on Education and Development of Civic Competencies, Seoul, October 2010).
2. U.S. Global Change Research Program, *Change Climate Literacy. The Essential Principles of Climate Science: A Guide for Individuals and Communities* U.S. Global Change Research Program, Washington, D.C., 2009), https://downloads.globalchange.gov/Literacy/climate_literacy_highres_english.pdf.

3. Christina Kwauk and Rebecca Winthrop, *Unleashing the Creativity of Teachers and Students to Combat Climate Change: An Opportunity of Global Leadership* (report, Brookings Institution, Washington, D.C., March 26, 2021), https://www.brookings.edu/research/unleashing-the-creativity-of-teachers-and-students-to-combat-climate-change-an-opportunity-for-global-leadership/.
4. Christina Kwauk and Amanda Braga, *Translating Competencies to Empowered Action. A Framework for Linking Girls' Life Skills Education to Social Change. Skills for a Changing World* (report, Brookings Institution, Washington, D.C., November 2017), https://www.brookings.edu/wp-content/uploads/2017/11/translating-competencies-empowered-action.pdf.
5. World Bank, *World Development Report 2018: Learning to Realize Education's Promise* (report, World Bank, Washington, D.C., 2018)https://www.worldbank.org/en/publication/wdr2018.
6. Joao Pedro Azevedo, "Learning Poverty: Measures and Simulations" (Policy Research Working Paper No. 9446, World Bank, Washington, D.C., 2020), https://openknowledge.worldbank.org/handle/10986/34654.
7. Christina Kwauk and Olivia Casey, *A New Green Learning Agenda: Approaches to Quality Education for Climate Action* (report, Center for Universal Education, Brookings Institution, Washington, D.C., 2021), https://www.brookings.edu/wp-content/uploads/2021/01/Brookings-Green-Learning-FINAL.pdf.
8. Kwauk and Casey, *A New Green Learning Agenda.*
9. The SDG 4 indicator list is at "SDG 4 Indicators," UNESCO, https://www.education-progress.org/en/indicators.
10. A. Gupta, M. Kalaivani, S. K. Gupta, S. K, Rai, and B. Nongkynrih, "The Study on Achievement of Motor Milestones and Associated Factors Among Children in Rural North India," *Journal of Family Medicine and Primary Care* 5 (April–June 2016): 378–82, https://journals.lww.com/jfmpc/Fulltext/2016/05020/The_study_on_achievement_of_motor_milestones_and.31.aspx.
11. World Bank, *World Development Report 2018.*
12. United Nations Educational, Scientific, and Cultural Organization, *Education for Sustainable Development Goals: Learning Objectives* (report, UNESCO, Paris, 2017), https://www.unesco.de/sites/default/files/2018-08/unesco_education_for_sustainable_development_goals.pdf.

13. R Iyengar and T. Stafford Ocansey, "Re-Imagining Schools to Support Psychosocial Well-Being of Teachers and Students as a Foundation for Effective Teaching and Learning During COVID-19 and Beyond," Global Happiness Council. Thematic Group: Education and Well-being (Policy Brief 1), 2021, https://s3.amazonaws.com/happinesscouncil.org/PB1_Education.pdf.
14. R. Winthrop, "The Need for Civic Education in the 21st Century Schools" (Policy Brief), June 4, 2020, https://www.brookings.edu/policy2020/bigideas/the-need-for-civic-education-in-21st-century-schools/.
15. See more details at State of New Jersey Department of Education, "New Jersey Student Learning Standards—Social Studies," https://www.state.nj.us/education/cccs/2020/2020%20NJSLS-SS.pdf, accessed May 1, 2023.

3. STRATEGIES IN INSTRUCTIONAL DESIGN

1. Intergovernmental Panel on Climate Change, "Global Warming of 1.5°C" (IPCC, 2019), https://www.ipcc.ch/site/assets/uploads/sites/2/2019/06/SR15_Summary_Volume_Low_Res.pdf.
2. Christina Kwauk and Olivia Casey, *A New Green Learning Agenda: Approaches to Quality Education for Climate Action* (report, Center for Universal Education, Brookings Institution, Washington, D.C., 2021), https://www.brookings.edu/wp-content/uploads/2021/01/Brookings-Green-Learning-FINAL.pdf.
3. F. Vona, G. Marin, D. Consoli, and D. Popp, *Green Skills* (NBER Working Paper 21116, National Bureau of Economic Research, Cambridge, MA, April 2015), http://www.nber.org/papers/w21116.
4. Christina Kwauk and Amanda Braga, *Translating Competencies to Empowered Action: A Framework for Linking Girls' Life Skills Education to Social Change* (report, Brookings Institution, Washington, D.C., November 2017), https://www.brookings.edu/wp-content/uploads/2017/11/translating-competencies-empowered-action.pdf.
5. Sasha Barab and Kurt Squire, "Design-Based Research: Putting a Stake in the Ground," *Journal of the Learning Sciences* 3, no. 1 (2004): 1–14.
6. David E. Kanter, "Doing the Project and Learning the Content: Designing Project-Based Science Curricula for Meaningful Understanding," *Science Education* 94, no. 3 (2020): 525–51, https://doi.org/10.1002/sce.20381.

7. Kanter, "Doing the Project and Learning the Content."
8. Elizabeth G. Cohen and Rachel Lotan, "Producing Equal-Status Interaction in the Heterogeneous Classroom," *American Educational Research Journal* 32, no. 1 (1995): 99–120, https://doi.org/10.3102/00028312032001099.
9. Cohen and Lotan, "Producing Equal-Status Interaction in the Heterogeneous Classroom."
10. Mika Munakata and Ashwin Vaidya, "Using Project- and Theme-Based Learning to Encourage Creativity in Science," *Journal of College Science Teaching* 45, no. 2 (2015): 48–53, http://www.jstor.org/stable/43631904.
11. Munakata and Vaidya, "Using Project- and Theme-Based Learning to Encourage Creativity in Science."
12. M. Pedaste, M. Mäeots, L. Siiman, T. de Jong, et al., "Phases of Inquiry-Based Learning: Definitions and the Inquiry Cycle," *Educational Research Review* 14 (2015): 47–61.
13. Pedaste et al., "Phases of Inquiry-Based Learning."
14. M. Panasan and P. Nuangchalerm, "Learning Outcomes of Project-Based and Inquiry-Based Learning Activities," *Journal of Social Sciences* 6, no. 2 (2010): 252–55.
15. National Youth Leadership Council, *K–12 Serving Learning Standards for Quality Practice* (report, National Youth Leadership Council, St. Paul, MN, 2008), https://cdn.ymaws.com/www.nylc.org/resource/resmgr/resources/lift/standards_document_mar2015up.pdf.
16. Alexander W. Astin, Lori J. Vogelgesang, Elaine K. Ikeda, and Jennifer A. Yee, "How Service Learning Affects Students," *Higher Education* 144 (2000), http://digitalcommons.unomaha.edu/slcehighered/144.
17. Astin et al., "How Service Learning Affects Students."
18. Betsy DiSalvo, Mark Guzdial, Amy Bruckman, and Tom McKlin, "Saving Face While Geeking Out: Video Game Testing as a Justification for Learning Computer Science," *Journal of the Learning Sciences* 23, no. 3 (2014): 272–315, https://doi.org/10.1080/10508406.2014.893434
19. Cinda Sue G. Davis and Cynthia Finelli, "Diversity and Retention in Engineering," *New Directions for Teaching and Learning* 3 (2007): 63–71.

4. CLIMATE CHANGE IN FORMAL LEARNING ENVIRONMENTS

1. National Research Council, "A Framework for K-12 Science Education: Practices, Crosscutting Concepts, and Core Ideas" (Washington, D.C.: National Academies Press, 2012), https://doi.org/10.17226/13165.
2. United Nations Educational, Scientific and Cultural Organization, *Education for Sustainable Development Goals: Learning Objectives* (report, UNESCO, Paris, 2017), https://www.unesco.de/sites/default/files/2018-08/unesco_education_for_sustainable_development_goals.pdf.
3. David Selby, "Review Article, *Creating Futures: 10 Lessons Inspiring Inquiry, Creativity, and Cooperation in Response to Climate Change for Senior Primary Classrooms*," *Policy and Practice: A Development Education Review* 25 (Autumn 2016): 153–60.
4. Selby, *Creating Futures*.
5. Rowan Oberman, *Creating Futures: 10 Lessons Inspiring Inquiry, Creativity, and Cooperation in Response to Climate Change for Senior Primary Classrooms* (Dublin: Education for a Just World [Trócaire/Centre for Human Rights and Citizenship Education, Dublin City University Institute of Education], 2016).
6. Selby, "Review Article: *Creating Futures*."
7. SHOREline, "SHOREline High School Students from New York City and the Gulf Coast Engage in a 'Katrina/Sandy Youth Dialogue'" (press release), https://bit.ly/40O05t8.
8. Columbia Climate School, National Center for Disaster Preparedness, GCPI Homehttps://ncdp.columbia.edu/microsite-page/gcpi/gcpi-home/.
9. See at State of New Jersey Department of Education, "New Jersey Student Learning Standards," https://www.nj.gov/education/standards/climate/learning/gradeband/index.shtml.
10. State of New Jersey Department of Education, "New Jersey Student Learning Standards, Science," https://www.nj.gov/education/standards/science/Index.shtml.
11. State of New Jersey Department of Education, "New Jersey Student Learning Standards, Science."

12. See State of New Jersey Department of Education, "New Jersey Student Learning Standards for Science," https://www.nj.gov/education/cccs/2020/.
13. See Green Schools Project, https://www.greenschoolsproject.org.uk/.
14. Zero Carbon Schools, https://www.greenschoolsproject.org.uk/zero-carbon-schools/.
15. N. Slawson, "Pupil Power: How Students Are Turning Schools Green," *Guardian*, April 28, 2017, https://www.theguardian.com/teacher-network/2017/apr/28/pupil-power-how-students-are-turning-schools-green.
17. U.S. Forest Service, Climate Science Resource Center, "Climate Science Primer," https://www.fs.usda.gov/ccrc/education/climate-primer. See also A. J. Hawkins and L. A. Stark, "Bringing Climate Change Into the Life Science Classroom: Essentials, Impacts on Life, and Addressing Misconceptions," *CBE—Life Sciences Education* 15, no. 2 (2016): fe3, https://doi.org/10.1187/cbe.16-03-0136.
18. Hawkins and Stark, "Bringing Climate Change into the Life Science Classroom."
19. NASA, Global Climate Change: Vital Signs of the Planet, https://climate.nasa.gov/.
20. Hawkins and Stark, "Bringing Climate Change into the Life Science Classroom."
21. Hawkins and Stark, "Bringing Climate Change into the Life Science Classroom."
22. A. F. Johnson et al., *Ecological Impacts of Climate Change* (Washington, D.C.: National Academies Press, 2009), https://doi.org/10.17226/12491. See also Hawkins and Stark, "Bringing Climate Change into the Life Science Classroom."
23. Johnson et al., *Ecological Impacts of Climate Change.*
24. See full details at National Park Service, "The Science of Climate Change in National Parks Video Series," http://www.nps.gov/subjects/climatechange/sciencevideos.htm. See also Hawkins and Stark, "Bringing Climate Change into the Life Science Classroom."
25. Hawkins and Stark, "Bringing Climate Change into the Life Science Classroom."
26. Committed to Climate and Energy Education, CLEAN Collection of Educational Resources, https://cleanet.org/clean/educational_resources/index.html. See full details at http://cleanet.org.

27. Hawkins and Stark, "Bringing Climate Change into the Life Science Classroom."
28. Hawkins and Stark, "Bringing Climate Change into the Life Science Classroom."
29. See full details at Clime Time: Climate Science Learning, https://www.climetime.org/.
30. Skeptical Science, https://www.skepticalscience.com. See also Hawkins and Stark, "Bringing Climate Change into the Life Science Classroom."
31. Hawkins and Stark, "Bringing Climate Change into the Life Science Classroom."
32. TROP ICSU, Teaching Resources to Integrate Climate Topics Across the Curricula, Across the Globe, https://tropicsu.org.
33. Green Ninja, https://web.greenninja.org.
34. SDGs Today, "Timely Data for Sustainable Development," https://sdgstoday.org/.
35. Design for Change, https://dfcworld.org/.
36. S. S. Siperstein, "Climate Change in Literature and Culture: Conversion, Speculation, Education" (PhD diss., University of Oregon, 2016), https://core.ac.uk/download/pdf/80854144.pdf.
37. See course details at Udemy, "SDG 4.7 Across Curriculum and Education Spaces," https://www.udemy.com/course/sdg47-across-curriculum/.
38. Siperstein, "Climate Change in Literature and Culture."
39. Siperstein, "Climate Change in Literature and Culture."
40. Siperstein, "Climate Change in Literature and Culture."
41. S. A. Ambrose, M. W. Bridges, M. DiPietro, M. C. Lovett, and M. K. Norman, *How Learning Works: Seven Research-Based Principles for Smart Teaching* (San Francisco: Jossey-Bass, 2010).

5. COMMUNITY-BASED (INFORMAL) EDUCATION

1. Patrick Werquin, *Terms, Concepts, and Models for Analysing the Value of Recognition Programs* (Vienna: Organization for Economic Co-operation and Development, 2007), https://www.oecd.org/education/skills-beyond-school/41834711.pdf.
2. M. Llopart and M. Esteban-Guitart, "Funds of Knowledge in 21st Century Societies: Inclusive Educational Practices for Under-Represented

Students. A Literature Review," *Journal of Curriculum Studies* 50, no. 2 (2018): 145–61.

3. L. C. Moll, C. Amanti, D. Neff, and N. Gonzalez, "Funds of Knowledge for Teaching: Using a Qualitative Approach to Connect Homes and Classrooms," *Theory into Practice* 31, no. 2 (1992): 133.
4. M. Esteban-Guitart and L. C. Moll, "Funds of Identity: A New Concept Based on the Funds of Knowledge Approach," *Culture & Psychology* 20, no. 1 (2014): 31–48, https://doi.org/10.1177/1354067X13515934.
5. Esteban-Guitart and Moll, "Funds of Identity."
6. J. M. Kiyama, "Family Lessons and Funds of Knowledge: College-Going Paths in Mexican American Families," *Journal of Latinos and Education* 10, no. 1 (2011): 23–42.
7. Moll et al., "Funds of Knowledge for Teaching."
8. Moll et al., "Funds of Knowledge for Teaching."
9. O. Lee A. Luykx, *Science education and student diversity: Synthesis and research agenda* (Cambridge: Cambridge University Press, 2006).
10. R. Zidny, J. Sjöström, and I. Eilks, "A Multi-Perspective Reflection on How Indigenous Knowledge and Related Ideas Can Improve Science Education for Sustainability," *Science & Education* 29 (2020): 145–85.
11. Zidny et al., "A Multi-Perspective Reflection on How Indigenous Knowledge and Related Ideas Can Improve Science Education for Sustainability."
12. F. Mazzocchi, "A Deeper Meaning of Sustainability: Insights from Indigenous Knowledge," *Anthropocene Review* 7, no. 1 (2020): 77–93.
13. A. Dixon, "Wetland Sustainability and the Evolution of Indigenous Knowledge in Ethiopia," *Geographical Journal* 171, no. 4 (2005): 306–23.
14. A. C. Obiora and E. E. Emeka, "African Indigenous Knowledge System and Environmental Sustainability," *International Journal of Environmental Protection and Policy* 3, no. 4 (2015): 88–96.
15. Obiora, "African Indigenous Knowledge System and Environmental Sustainability."
16. L. Ramirez, "The Importance of Climate Education in a COVID-19 World," World Economic Forum, May 8, 2020, https://www.weforum.org/agenda/2020/05/the-importance-of-climate-education-in-a-covid-19-world/.
17. C. Kwauk and R. Winthrop, *Unleashing the Creativity of Teachers and Students to Combat Climate Change: An Opportunity of Global Leadership*

(report, Brookings Institution, Washington, D.C., March 26, 2021), https://www.brookings.edu/research/unleashing-the-creativity-of-teachers-and-students-to-combat-climate-change-an-opportunity-for-global-leadership/.

18. R. Blanco and M. Umayahara, *Participación de las familias en la educación infantil latinoamericana* [Participation of families in Latin American early childhood education] (Santiago, Chile: UNESCO, Oficina regional de educación para América Latina y el Caribe, 2004).
19. Ramirez, "The Importance of Climate Education in a COVID-19 World."
20. Trash Hack, https://www.trashhack.org/the-green-future-is-here-in-bhopal/.
21. A. Shtivelband, A. W. Roberts, and R. Akubowski, *STEM Equity in Informal Learning Settings: The Role of Libraries* (Datum Advisors, December 2016), http://ncil.spacescience.org/images/papers/STEM Equity in Informal Learning Settings_FINAL.pdf.
22. Shtivelband et al., *STEM Equity in Informal Learning Settings*.
23. Shtivelband et al., *STEM Equity in Informal Learning Settings*.
24. American Library Association (ALA), "STAR Net STEAM Equity Project: Enhancing Learning Opportunities in Libraries of Rural Communities," https://www.ala.org/tools/programming/steamequity.
25. ALA, "STAR Net STEAM Equity Project."
26. ALA, "STAR Net STEAM Equity Project."
27. Shtivelband et al., *STEM Equity in Informal Learning Settings*.
28. Shaky Sherpa, "Shared Solar: Prepaid Electricity via Mobile Telephony," Quadracci Sustainable Engineering Lab, Columbia Climate School, September 14, 2011, https://qsel.columbia.edu/shared-solar-2/.
29. R. Cho, "GRID3 Project Aims to Put Everyone on the Map," State of the Planet, Columbia Climate School, December 13, 2019, https://news.climate.columbia.edu/2019/12/04/grid3-population-mapping/.
30. V. Friedman, "The Future Is Trashion," *New York Times*, December 26, 2019, https://www.nytimes.com/2019/12/20/style/zero-waste-daniel-trashion.html.
31. R. I. Iyengar, "Day 2: Pushing for Climate Action Inside and Outside the Classroom," February 10, 2021, https://blogs.cuit.columbia.edu/edsdcsd/2021/02/10/day-2-pushing-for-climate-action-inside-and-outside-the-classroom/

32. T. Ocansey Stafford and E. Siakwa Nuetey, "Eco-Conscious Community Development in Non-Formal Education," in *Curriculum and Learning for Climate Action: Toward an SDG4.7 Roadmap for Systems Change*, IBE Series on Curriculum, Learning, and Assessment, vol. 5, ed. Radhika Iyengar and Christina Kwauk (Leiden, Netherlands: Brill, 2021).
33. D. Resnik, K. Elliott, and A. Miller, "A Framework for Addressing Ethical Issues in Citizen Science," *Environmental Science & Policy* 54 (20150: 476.
34. R. Iyengar, "Glimpses Day 1" (video), YouTube, November 13, 2019, https://www.youtube.com/watch?v=PDDiEDe8zWU.
35. R. Iyengar, "Explaining the Fluorosis Problem Using Pictorial Posters" (video), YouTube, September 9, 2019, https://www.youtube.com/watch?v=jAPoZAD3yYQ.
36. Iyengar, "Explaining the Fluorosis Problem Using Pictorial Posters."
37. R. Iyengar, "Education Brings Sectors Together to Address Fluorosis in Alirajpur," State of the Planet, Columbia Climate School, August 12, 2019, https://news.climate.columbia.edu/2019/08/12/education-interventions-high-fluoride-india/.
38. R. Iyengar, "Women in Science" (video), YouTube, September 9, 2019, https://www.youtube.com/watch?v=QMhJjbobX9Y.
39. R. Iyengar, "Street Theatre to Make People Aware About Fluorosis" (video), YouTube, September 9, 2019 https://www.youtube.com/watch?v=lKQBCDOi-D8.
40. R. Iyengar, "Process Documentation of Testing for Fluoride in Alirajpur" (video), YouTube, September 16, 2019, https://www.youtube.com/watch?v=6eW3Vi9XOEc.

6. TEACHING CLIMATE CHANGE IN NONFORMAL SETTINGS

1. Council of Europe, Linguistic Integration of Adult Migrants (LIAM), "Formal, Non-Formal, and Informal Learning," https://www.coe.int/en/web/lang-migrants/formal-non-formal-and-informal-learning.
2. Standard University, Doerr School of Sustainability, "Q&A: Climate Change Politics in the High School Classroom," https://earth.stanford.edu/news/qa-climate-change-politics-high-school-classroom.

3. Columbia Climate School, Lamont-Doherty Earth Observatory, "Next Generation of Hudson River Educators," https://lamont.columbia.edu/ldeo-hudson-river-field-station/summer-high-school-opportunities.
4. Columbia Climate School, Lamont-Doherty Earth Observatory, "Next Generation of Hudson River Educators 2020, " https://lamont.columbia.edu/ldeo-hudson-river-field-station/summer-high-school-opportunities/next-gen-2020.
5. NYC Science Research Mentoring Consortium, https://www.studentresearchnyc.org/.
6. NYC Science Research Mentoring Consortium, Woodland Ecology Research Mentorship (WERM), https://www.studentresearchnyc.org/our_programs/woodland-ecology-research-mentorship-werm/.
7. Wave Hill, "14 Months as a WERM," https://www.wavehill.org/education/youth-internships/werm/werm-phases.
8. Community for Advancing Discovery Research in Education, "LabVenture—Revealing Systemic Impacts of a 12-Year Statewide Science Field Trip Program"(Education Development Center, Boston, MA, 2021), http://cadrek12.org/projects/labventure-revealing-systemic-impacts-12-year-statewide-science-field-trip-program.
9. Community for Advancing Discovery Research in Education, "LabVenture."
10. GreenHub: Youth Fellowship and Video for Change, https://www.greenhubindia.net/.
11. S. S. Siperstein, "Climate Change in Literature and Culture: Conversion, Speculation, Education" (PhD diss., University of Oregon, 2016), https://core.ac.uk/download/pdf/80854144.pdf.
12. Siperstein, "Climate Change in Literature and Culture."
13. Siperstein, "Climate Change in Literature and Culture."
14. Gorongosa National Park, "Girls Clubs," https://gorongosa.org/girls-clubs/.
15. D. Tham, "The Women Behind the Comeback: How One of Africa's National Parks Is Thriving After War," CNN, July 3, 2020, https://www.cnn.com/travel/article/inside-africa-gorongosa-national-park-spc-intl/index.html.
16. Gorongosa National Park, "Girls Clubs."
17. D. Gonçalves, "Mozambique's Trailblazing Gorongosa Park Celebrates 60th Anniversary, Announces 60 New Schools" (commentary), Mongabay,

July 23, 2020, https://news.mongabay.com/2020/07/mozambiques-trailblazing-gorongosa-park-celebrates-60th-anniversary-commentary/.

7. DIVERSITY, EQUITY, INCLUSION, AND ACCESS AS A TOOL TO ADDRESS SOCIAL AND ENVIRONMENTAL JUSTICE

1. Shirley Rainey and Glenn Johnson, "Grassroots Activism: An Exploration of Women of Color's Role in the Environmental Justice Movement," *Race, Gender, and Class* 16, nos. 3/4 (2009): 144–73. See also Ann Kushmerick, Lindsay Young, and Susan Stein, "Environmental Justice Content in Mainstream US, 6–12 Environmental Education Guides," *Environmental Education Research* 13, no. 3 (2007): 385–408.
2. A. Shtivelband, A. W. Roberts, and R. Akubowski, *STEM Equity in Informal Learning Settings: The Role of Libraries* (Datum Advisors, December 2016), http://ncil.spacescience.org/images/papers/STEM Equity in Informal Learning Settings_FINAL.pdf.
3. A. Lynch and J. Sachs, *United States Sustainable Development Report 2021* (SDSN, New York, November 19, 2021), https://us-states.sdgindex.org/.
4. R. Iyengar and M. Bajaj, "After the Smoke Clears: Toward Education for Sustainable Development in Bhopal, India," *Comparative Education Review* 55, no. 3 (2011).
5. Kushmerick, Young, and Stein, "Environmental Justice Content."
6. S. Lewis and K. James, "Whose Voice Sets the Agenda for Environmental Education? Misconceptions Inhibiting Racial and Cultural Diversity," *Journal of Environmental Education* 26, no. 3 (1995).

INDEX

Italicized page numbers indicate figures or tables.